AF787043

Nanocellulose - Sources, Preparations, and Applications

Edited by Md. Salim Newaz Kazi

Published in London, United Kingdom

Nanocellulose - Sources, Preparations, and Applications
http://dx.doi.org/10.5772/intechopen.111178
Edited by Md. Salim Newaz Kazi

Contributors
Anwesha Barik, Caroliny M. Santos, Elham Asghari-Varzaneh, Gara Dheeraj Kumar, Haiyong Ao, Hajar Shekarchizadeh, Hanumanthu Jeevan Rao, Jinlong Zhang, Kaleemullah Shaikh, Kirubanandan Shanmugam, Md. Salim Newaz Kazi, Mohd Nashrul Mohd Zubir, Moumita Das Kirtania, Nandini Robin Nadar, Narender Singh, Perumalla Janaki Ramulu, Qinglin Wu, Sanjay Dey, Sanjay Singh, Subhasish Chakraborty, Thiago F. Santos, Wajahat Ahmed Khan, Weiguo Li, Xiaowei Xun

© The Editor(s) and the Author(s) 2024
The rights of the editor(s) and the author(s) have been asserted in accordance with the Copyright, Designs and Patents Act 1988. All rights to the book as a whole are reserved by INTECHOPEN LIMITED. The book as a whole (compilation) cannot be reproduced, distributed or used for commercial or non-commercial purposes without INTECHOPEN LIMITED's written permission. Enquiries concerning the use of the book should be directed to INTECHOPEN LIMITED rights and permissions department (permissions@intechopen.com).
Violations are liable to prosecution under the governing Copyright Law.

Individual chapters of this publication are distributed under the terms of the Creative Commons Attribution 3.0 Unported License which permits commercial use, distribution and reproduction of the individual chapters, provided the original author(s) and source publication are appropriately acknowledged. If so indicated, certain images may not be included under the Creative Commons license. In such cases users will need to obtain permission from the license holder to reproduce the material. More details and guidelines concerning content reuse and adaptation can be found at http://www.intechopen.com/copyright-policy.html.

Notice
Statements and opinions expressed in the chapters are these of the individual contributors and not necessarily those of the editors or publisher. No responsibility is accepted for the accuracy of information contained in the published chapters. The publisher assumes no responsibility for any damage or injury to persons or property arising out of the use of any materials, instructions, methods or ideas contained in the book.

First published in London, United Kingdom, 2024 by IntechOpen
IntechOpen is the global imprint of INTECHOPEN LIMITED, registered in England and Wales, registration number: 11086078, 167-169 Great Portland Street, London, W1W 5PF, United Kingdom

British Library Cataloguing-in-Publication Data
A catalogue record for this book is available from the British Library

Additional hard and PDF copies can be obtained from orders@intechopen.com

Nanocellulose - Sources, Preparations, and Applications
Edited by Md. Salim Newaz Kazi
p. cm.

This title is part of the Nanotechnology and Nanomaterials Book Series, Volume 3
Topic: Nanomaterials and Nanostructures
Series Editor: Jung Huang
Topic Editor: Alberto Jiménez-Suárez
Associate Topic Editor: Gonzalo Seisdedos

Print ISBN 978-1-83769-258-3
Online ISBN 978-1-83769-259-0
eBook (PDF) ISBN 978-1-83769-260-6
ISSN 3029-0538

We are IntechOpen, the world's leading publisher of Open Access books

Built by scientists, for scientists

7,000+
Open access books available

187,000+
International authors and editors

205M+
Downloads

156
Countries delivered to

Our authors are among the
Top 1%
most cited scientists

12.2%
Contributors from top 500 universities

WEB OF SCIENCE™

Selection of our books indexed in the Book Citation Index in Web of Science™ Core Collection (BKCI)

Interested in publishing with us?
Contact book.department@intechopen.com

Numbers displayed above are based on latest data collected.
For more information visit www.intechopen.com

For EU product safety concerns:
IN TECH d.o.o., Prolaz Marije Krucifikse Kozulić 3, 51000 Rijeka, Croatia,
info@intechopen.com or visit our website at intechopen.com.

IntechOpen Book Series

Nanotechnology and Nanomaterials

Volume 3

Aims and Scope of the Series

Humans face growing challenges in environmental sustainability. We need better materials, more effective devices, and higher computing power to conquer these challenges. Researchers have long explored the realm of science at the nanometer scale for developing novel problem-solving technologies. In the past decades, breakthroughs in multiscale computer simulation, novel hierarchical material and device structures, nanometrology, and new functionalities from interfacial and quantum size effects have enhanced our ability to utilize nanotechnology to solve real-world problems. This book series addresses important advancements in nanotechnology and nanomaterials. It showcases how nanotechnology is being continually developed and implemented in a variety of domains of science and technology. The two main topics this series covers are Nanotechnology and Nanodevices, and Nanomaterials and Nanostructures.

Meet the Series Editor

Jung Y. Huang, a university educator and researcher, has been working to unravel the structures and functional properties of materials and cellular events in living cells with varying optical methodologies. He has co-authored hundreds of journal papers and five book chapters. He also holds tens of patents in laser techniques and single-molecule/hyperspectral imaging and has developed architectural photonics based on hierarchically structured materials. Currently, his research focuses on the integration of artificial intelligence methodology with optics to automatically discover meaningful information from optical sensing/imaging data cubes. He has been an editorial board member and reviewer for several scientific journals. As a member of the global scientific community, he sincerely supports and endeavors to promote the spread of scientific knowledge.

Meet the Volume Editor

Dr. Salim Newaz Kazi is a Professor of Mechanical Engineering at Universiti Malaya, Malaysia. He obtained his BSc, MSc, and MS in Mechanical Engineering and his Ph.D. in Chemical and Materials Engineering. After a long career in the petrochemical industry and consultancy work in different engineering companies, Dr. Kazi transitioned to academics in 2009. He is specialized in heat transfer, fluid mechanics, particle characterization, heat exchanger fouling mitigation, nanofluid synthesis and applications, and renewable energy. He has many years of engineering service experience in petrochemical industries and chemical plants where he worked as a senior engineer and plant manager. He also worked as a consultant for different engineering companies and was involved in plant design and installation, troubleshooting, maintenance, and so on. He is a Chartered Engineer (CEng, UK), a fellow of the Institute of Engineers Bangladesh (FIEB, Bangladesh), and a member of several professional bodies, including the Institution of Mechanical Engineering (IMechE, UK) and the Board of Engineers Malaysia (BEM, Malaysia). Dr. Kazi has supervised and evaluated many Ph.D. and master's theses. He has published numerous technical papers in national and international journals and conference proceedings. He has also published technical book chapters and edited engineering books. He has participated in many international engineering conferences and congresses and has given invited speeches. He has also filed many patents. Dr. Kazi received several awards in recognition of his contributions to research and innovation. He is an active member of the editorial board of several international journals. Dr. Kazi has developed an active research network both at home and abroad.

Contents

Preface

Nanocellulose derived from natural cellulosic materials has attracted significant interest due to its mechanical and optical properties, as well as its eco-friendliness, wide availability, recyclability, renewability, and low coefficient of thermal expansion (CTE). Various extraction techniques are employed to obtain nanocellulose, each with distinct advantages. Mechanical methods such as high-pressure homogenization and ultrasonication break down cellulose fibers effectively into nanoscale dimensions. Conversely, chemical approaches like acid hydrolysis and oxidation selectively remove amorphous portions, resulting in highly crystalline nanocellulose. Additionally, environmentally friendly extraction methods involve enzymatic processes that utilize cellulase enzymes to extract nanocellulose from cellulose-rich sources.

In an era marked by increasing environmental concerns and the quest for sustainable solutions, nanocellulose emerges as an inspiration of hope. This book, *Nanocellulose – Sources, Preparations, and Applications,* takes readers on a journey through the world of this remarkable material. Edited with meticulous care, this comprehensive volume stands as a testament to the collective efforts of researchers pushing the boundaries of material science.

The utilization of free-standing cellulose nanofiber (CNF) film extends to various applications, including serving as a potential barrier, packaging material, and membranes for wastewater treatment, as well as for fabricating biomedical films for wound repair, among other uses. The spraying process emerges as a notable method for large-scale production of films quickly and efficiently. The film's smooth and glossy surface offers opportunities for the development of functional materials, such as serving as a base substrate for flexible electronics and solar cells, leveraging its smoothness.

Nanocellulose has garnered significant attention due to its biocompatibility, renewability, robust mechanical strength, and cost-effectiveness, leading to exploration of its utilization across a wide array of applications. These applications include modifying nanocellulose for wettable applications, incorporating it as a carrier for antimicrobial agents, utilizing it as a barrier against UV rays or chemicals, and employing it for reinforcement in various fields such as biomedical, automotive, electronic, and energy materials. Despite promising potential, the utilization of nanocellulose remains an emerging field, encountering numerous technical challenges in achieving reliability, renewability, and sustainability for modern applications.

A promising source for ecofriendly biomaterial extraction is lignocellulosic natural fibers derived from trees, plants, agricultural waste, fruits, vegetables, seeds, and leaves. Various methods, including mechanical, chemical, and enzymatic procedures, are employed to extract nanocellulose from these natural sources. These techniques enable the production of nanocellulose with tailored morphologies, crystallinities, and surface functionalities, making them suitable for specific applications.

Advancements in nanocellulose preparation techniques involve the use of functional additives, surface modifications, and composite formations. These innovative approaches enhance the mechanical, thermal, and barrier properties of nanocellulose, thereby expanding its potential for use in cutting-edge applications.

The potential applications of the product span various research domains, including tissue engineering scaffolds and drug delivery systems within the biomedical sector, as well as the development of flexible and sustainable electronic devices in the realm of nanoelectronics. Nanocellulose serves as an advantageous component for enhancing the mechanical properties of composites due to its exceptional reinforcing capabilities in polymeric matrices. Its utilization extends to environmental applications such as water purification, oil spill clean-up, and the production of biodegradable packaging materials. Unlike cellulose sourced from plants and wood, bacterial-derived nanocellulose boasts the greatest purity and natural nanofiber morphology, garnering increasing interest from researchers and industries alike. It possesses several unique attributes, including biomimetic nanoscale three-dimensional network structure, high water retention capacity, moldability into various shapes, inherent wet strength, excellent gas permeability, and favorable biocompatibility.

Pesticides play a vital role in safeguarding agricultural produce, protecting stored grain commodities and field crops from insect pests. Nanomaterials applied in plant protection offer numerous advantages over traditional pesticide products. Nano-enabled pesticides have the potential to alter the functionality or risk profile of active ingredients. Advanced nanotechnology has led to the development of new pesticide formulations that are less environmentally detrimental, cost-effective, and more efficient compared to conventional formulations. These innovations hold the promise of reducing environmental pollutants and fostering sustainable economic growth.

A paradigm shift towards sustainable practices is occurring, and the significance of nanocellulose cannot be overstated. From its renewable sourcing to its biodegradability, nanocellulose embodies the principles of green chemistry, offering a viable alternative to conventional materials.

This book serves as a repository of knowledge, exploring the diverse sources and preparations of nanocellulose. Chapter authors illuminate the intricacies of its production, from isolation techniques to functionalization methods. By providing a comprehensive overview, the book serves as a useful reference for researchers and practitioners, providing them with the tools necessary to harness the full potential of this sensation material.

Moreover, *Nanocellulose – Sources, Preparations, and Applications* transcends mere academic discourse, offering insights into real-world applications. From biomedical engineering to sustainable packaging, nanocellulose demonstrates its versatility across numerous fields. By showcasing innovative case studies and cutting-edge research, the book inspires readers to envision novel solutions to contemporary challenges.

As we navigate the complexities of the current century, the urgency of sustainable innovation has never been more apparent. In this context, nanocellulose emerges as a game-changer, poised to revolutionize industries and redefine our relationship with the environment. Yet, realizing this vision necessitates collaboration, innovation, and a commitment to pushing the boundaries of scientific inquiry.

The book on nanocellulose represents a manifesto for a sustainable future. It encourages us to embrace the transformative potential of nanocellulose and embark on a journey towards a more sustainable and prosperous tomorrow.

Although this work was an addition to my routine academic workload, it was thoroughly enjoyable. I would like to thank my students for their patience in receiving delayed support from me. Lastly, I would like to thank my son Md Mehrab Newaz Kazi and my wife Nilufa Parveen for their sacrifices in our daily life throughout the duration of my editing work.

Salim Newaz Kazi
Department of Mechanical Engineering,
Faculty of Engineering,
Universiti Malaya,
Kuala Lumpur, Malaysia

Section 1

Fundamentals of Nanocellulose

Chapter 1

Nanocellulose: Fundamentals and Applications

Kaleemullah Shaikh, Wajahat Ahmed Khan, Md. Salim Newaz Kazi and Mohd Nashrul Mohd Zubir

Abstract

Cellulose is a natural and abundant polymer which can be derived from a large variety of materials such as biomass, plants and animals etc. Nanocellulose demonstrates remarkable physicochemical, mechanical, biological and structural properties. Technological challenges such as efficient extraction of cellulose and nanocellulose from precursors are still a challenge. Several techniques such as chemical, mechanical, biological, and combined approaches are utilized for the preparation of desired nanocellulose. However, the processes available to manufacture nanocellulose are still expensive. One of the most common methods used to obtain cellulose nanocrystals is acid hydrolysis method with strong acids such as sulfuric or hydrochloric acid. Recently nanocellulose has gained great attention due to their biocompatibility, renewable nature, mechanical strength, and cost-effectiveness. Hence wide range of applications for nanocellulose are being explored such as wettable applications to make hydrophobic modification for nanocellulose, or as a carrier of antimicrobial substances, or as creating a barrier from UV rays or from chemicals, it is also being used for reinforcement, biomedical, automobiles electronic, and energy materials. However, utilization of nanocellulose is still an emerging field and faces lots of technical challenges to be utilized as a reliable, renewable, and sustainable material for modern applications.

Keywords: nanocellulose, cellulose-nanocrystals (CNCs), automobiles, smart responsive device, corrosion protector

1. Introduction

One of the most abundant biopolymers on earth is cellulose, a substance identified by Payen in 1838. It is a polysaccharide composed of repeated anhydroglucose units (AGUs) connected jointly via β-1,4-glycoside linkages. Each cellulose unit consists of three hydroxyl groups, 1 primary and 2 secondaries, as shown in **Figure 1**. These groups and their hydrogen bonding capability, in same chain and different chain, with adjacent groups are crucial in regulating the significant physical characteristics and directing crystallization packing. Cellulose morphology is fibrous with irregular crystalline and amorphous segments, that is, they are built up of bundles/aggregates of fibrils where every fibril consists of large order repetitive (crystalline) regions and small amorphous regions. The crystalline regions contain properly packed chain

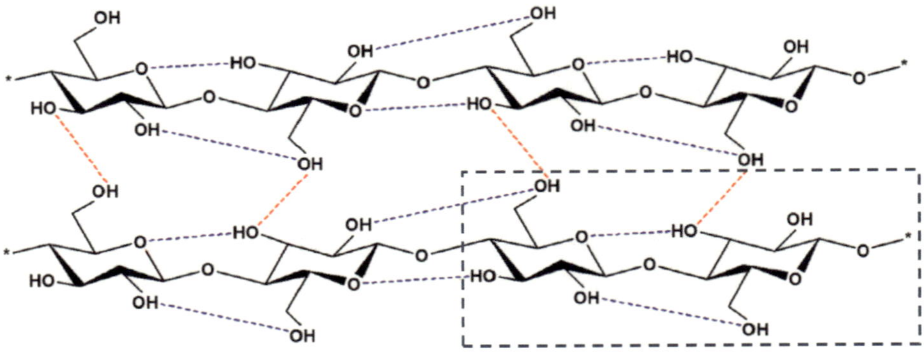

Figure 1.
Unit structure of cellulose, intramolecular and intermolecular hydrogen bonding networks in cellulose structure [1].

molecules which demonstrate greater strength and stiffness of cellulose, and the amorphous regions provide the flexibility of bulk material [1].

Cellulose, an organic polymer, has been utilized since hundreds of years as a fiber or its derived substances over a large range of materials and final product applications. Cotton is the purest form of organic cellulose, that includes ~90% cellulose, while wood contains around 40–50% cellulose. Additionally, the polymerization degree (PD) of cellulose chains contains around ~10,000 glucopyranose units in wood and ~ 15,000 in cotton cellulose [2]. The percentage variations in these celluloses relies on biomass type, soil, location, weather and yielding time. It is one of the key components in structural plants elements and living organisms supporting retain their form, also, the total amount of cellulose produced in world by all living organisms is around $10^{11} - 10^{12}$ tons/year [3]. Cellulose is found in conjunction with hemicelluloses, lignin, and various other minor constituents within lignocellulosic biomass. Cellulose fibers are found in plants, stems, and forests are arranged as a support structure within a matrix of lignin [4]. The majority of animal species could not produce cellulose, while humans cannot even digest it.

The single cellulose molecules chain are assembled into primary cellulose fibrils of around 3.0–4.0 nm width and a few μm in length, through the hydrogen bonds among the hydroxyl groups of the anhydro glucose repeating units [2, 5]. The primary fibrils accumulate through hydrogen bonding interactions with adjacent primary fibrils,

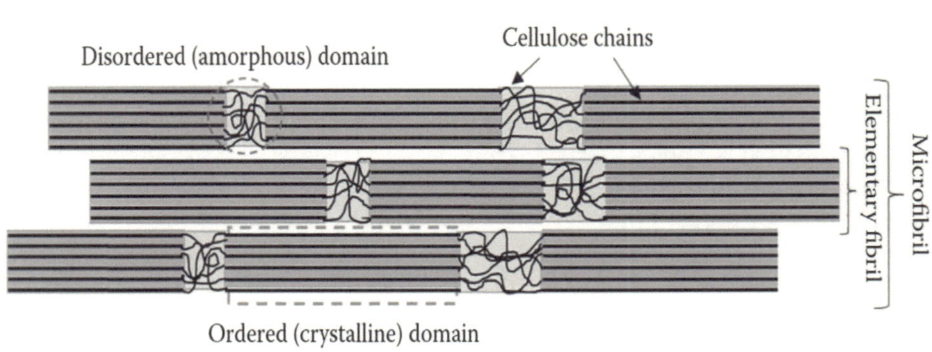

Figure 2.
Schematic of a cellulose microfibril (nanocellulose) structure showing both amorphous and crystalline regions [6].

generate bundles of nanostructure fiber called cellulose micro fibrils as demonstrated in **Figure 2**. Such microfibrils are sometimes referred to as nanocellulose or cellulosic nanoparticles.

2. Nanocellulose

Nanocellulose demonstrates remarkable physicochemical, mechanical, biological and structural properties. A wide range of applications for nanocellulose are presently being explored, including its potential usage in coatings, food sector, biomedical usage, hydrogels, treatment of wastewater, and applications in energy [7]. The utilization of nanomaterials has significantly risen during the last few years as a viable substitute for traditional artificial polymers. The polymers derived from nanocellulose have offered significant opportunities for both research and material production due to their inherent characteristics, including biodegradability, ease in processing, recycling capacity, and widespread availability [8, 9]. Furthermore, their low price, low density, reactive surface, capability to alter chemistry of surface, and remarkable mechanical characteristics, makes them suitable reinforcement material. Creating environmental sustainable and economical procedure for separating the nanocellulose from plant biomass is major achievement of real sustainability for getting full advantages from numerous applications of cellulose [10]. Nanocellulose comprises a significant quantity of hydroxyl (OH) groups, making it extremely hydrophilic and can be altered through various chemical and physical methodologies [4]. Therefore, because of the remarkable characteristics nanocellulose has gained enormous attention as an effective nanostructure to develop emerging nanomaterials (**Figure 3**) [11, 12].

Figure 3.
General properties of Nanocellulose [13].

2.1 Classification of nanocellulose

Cellulose is usually classified as per its source of origin, namely plant-based cellulose (PC), algae-based cellulose, wood-based cellulose (WC), animal-based cellulose (AC), and bacteria-based cellulose (BC). Moreover, nanocellulose can be classified based on its characteristics and functions or based on preparation methods and raw materials. The nanocellulose common classification consists of Nanocrystalline Cellulose (NCC)/ Cellulose-Nanocrystals (CNCs), Nano-fibrillated cellulose (NFC), Nano-whiskers (CNWs)/Cellulose-Nanofibrils (CNFs), and Microbial Cellulose/ Bacterial Cellulose (BCs). While all these are chemically similar, they demonstrate variation in size of particles, structure, crystallization, and other characteristics because of the particular sources and extraction procedures utilized. The **Table 1** below shows primary characteristics of nanocellulose types (**Figure 4**) [6, 13].

Material	Crystallinity	Surface Area (m^2g^{-1})	Youngs Modulus (GPa)	Average Size	
				Diameter	Length
Nanofibrillated Cellulose	50% ≈ 90%	≈ 100	50–160	10–80 nm	< 10 μm
Nanocrystalline Cellulose	≈ 90%	≈ 200	50–140	5–30 nm	≈ 100 nm
Bacterial Nanocellulose	> 80%	≈ 100	15–35	20–100 nm	1.0–5.0 μm

Table 1.
Primary characteristics of Nanocellulose types [13].

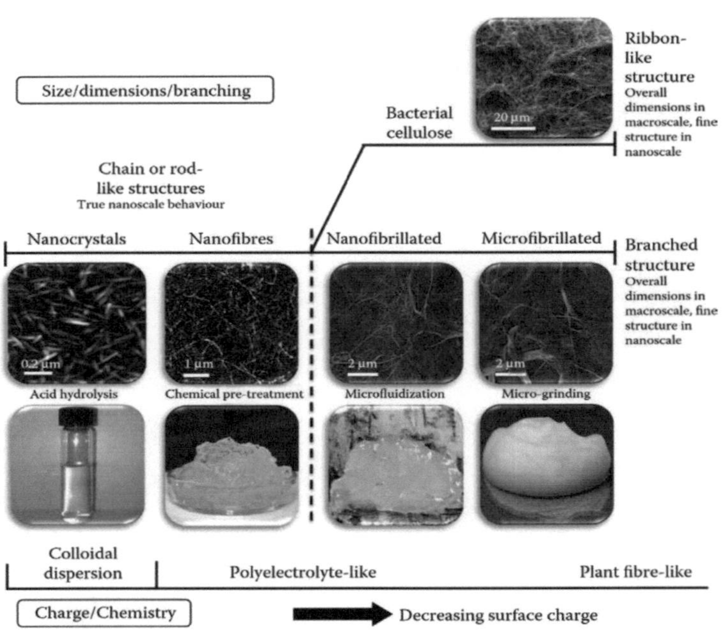

Figure 4.
Different types of nanocellulose materials categorized with respect to size and dimensions [6].

2.1.1 Nanocrystalline cellulose

As the name suggests these nanocellulose are made up of high crystallinity around 54–88%, and have shape like short rod, needle or whiskers. These are isolated crystallites having dimensions around 2–20 nm in diameter and 100–500 nm in length. These can be extracted from cellulose fibrils through various techniques, as shown in **Figure 5** [13].

Hydrolysis is a process where amorphous elements are hydrolyzed and eliminated through acid. However, the crystalline items remain without any change. The procedure contains 2 steps: raw material pretreatment preliminary treatment proceeded through its hydrolysis into cellulose nanocrystals (CNCs) [6]. The raw material is comprised of various impure substances, such as wax, esters, hemicelluloses, and lignin. These impurities can be eliminated through an alkaline (NaOH) solution treatment or by employing a bleaching technique just before the process of hydrolysis. After that cleaned raw material has been heat treated in the acidic environment about 45.0 minutes to many hours to hydrolyze the amorphous parts of fibers cellulose [4]. Several acidic minerals could be utilized for this process, that is, hydrochloric, maleic, sulfuric, phosphoric, hydrobromic formic and nitric acids. The acid hydrolysis of cellulose chains within amorphous domains includes immediate protonation of glucosidic oxygen (path 1) or cyclic oxygen (path 2), leading to gradual separation of the glucosidic bonds induced through the inclusion of water. The hydrolysis procedure produces 2 smaller chain parts, while maintaining the fundamental backbone structure [14]. Generally, it is observed that greater acid concentrations, prolonged reaction durations, and greater temperatures causes a greater surface charge and reduced particle sizes of cellulose nanocrystals (CNC). But these conditions lower the yield, as well as reduced crystallization and thermal durability of CNC (**Figure 6**) [14].

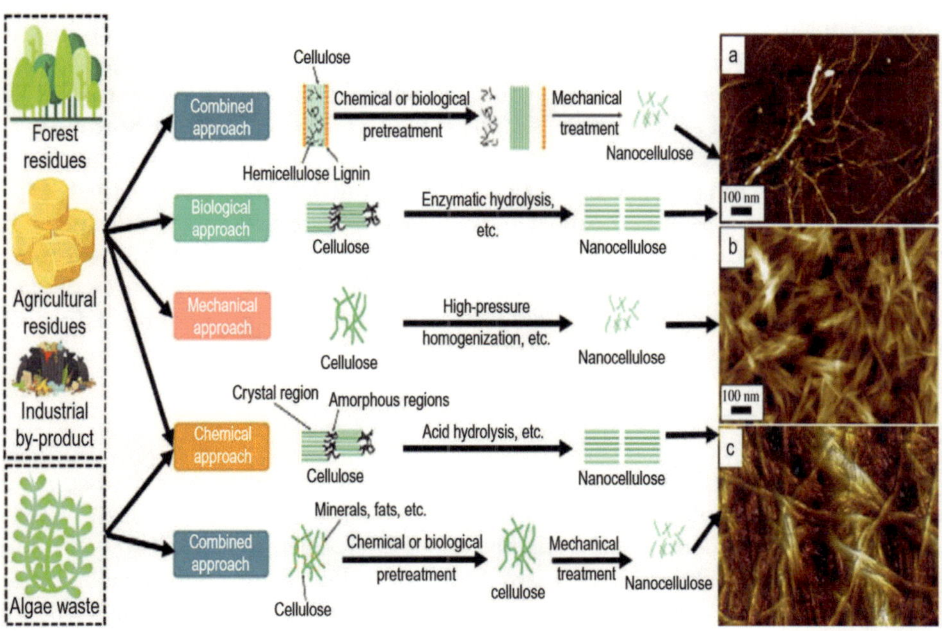

Figure 5.
Various preparation methods for Nanocellulose.

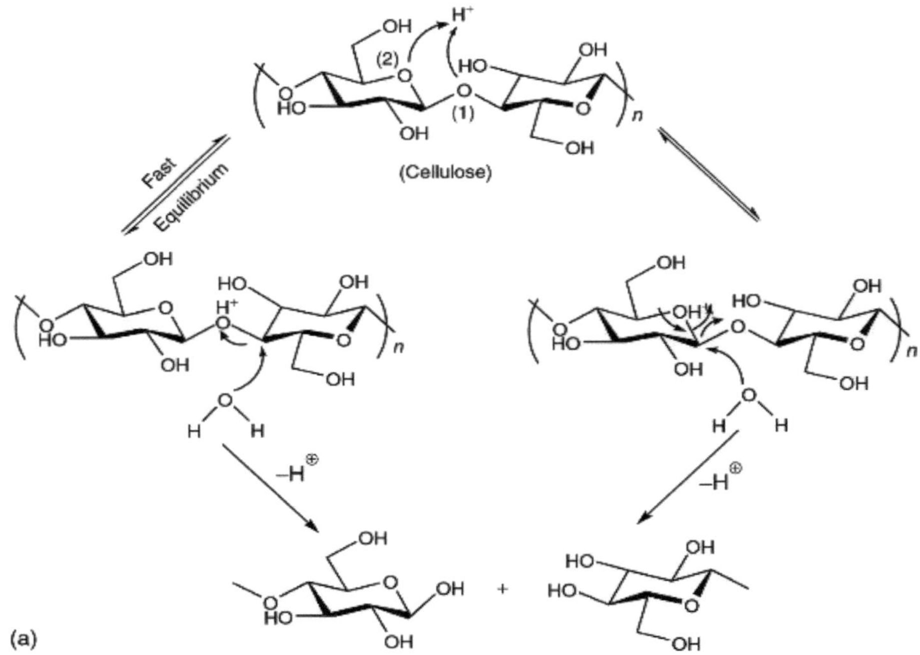

Figure 6.
Mechanism of cellulose chain acid hydrolysis [14].

The commonly used acid in this treatment is sulfuric acid because it efficiently degrades cellulose and also introduces sulphate half-esters on the surface of CNCs. These sulphates bear a monovalent charge, resulting in colloidal stability for the CNC dispersions in water due to electrostatic repulsion. However, it is an expensive method, consuming a lot of water whereby the possibility to completely recycle the acid is prevented. By and large, it is reasonable to state that the challenges posed by the sulfuric acid method have strongly impeded the industrial development of CNCs. Moreover, the yield of acid hydrolysis method is also rather low, generally around 30%. The yield of cellulose nanocrystals in the various treatments using conventional and microwave heating ranged between 3.4–29.0% and 4.9–38.2%, respectively [3].

Source material is seen as the most decisive factor when selecting the desired dimensions for CNCs. **Table 2** demonstrates the well-known effect of source materials on the width and length of the CNC. Generally, the width is thought to be determined by the microfibril width in the source, whereas the length is connected to its LODP value [15, 16].

2.1.2 Nanofibrils cellulose

Cellulose nanofibrils (CNFs) are defined as fibrillar structures with diameters ranging from a few hundred nanometers or less. Fibrils higher than dimensions are generally defined as micro fibrillated or cellulose microfibrils (CMF) [14].

Nano fibrillated cellulose (NFC), also known as cellulose microfibril, cellulose nanofibrillar cellulose, cellulose nanofiber or cellulose nanofibril is a highly flexible and entanglement nanocellulose that may be mechanically distinguished from cellulose fibrils [13]. When comparing nanocrystalline cellulose to nano fibrillated cellulose, it can be observed that later has higher length to diameter aspect ratio,

Cellulose Source	Dimensions (D: Diameter, L: Length, T: Thickness)
Sweet Potato	D: 20–40 nm
Wood Chips	L: 225–350 nm, D: 4.5–6 nm
Corn Husk	L: 20–254 nm, D: 0.8–11.8 nm
Sugarcane Bagasse	L: 37–220 nm, D: 18–32 nm
Banana Fibers	L: 135–145 nm, D: 7–10 nm
Plum Seed Shells	L: 100–800 nm
Bacteria	L: 160–420 nm, T: 15–25 nm
Cotton	L: 100–300 nm, T: 3–5 nm
Sisal	L: 100–300 nm, T: 3–5 nm
Ramie	L: 70–200 nm, T: 5–15 nm
Tunicates	L: 195–225 nm, T: 78–90 nm
Wood Pulp	L: 200–500 nm, D: 3–35 nm
Switchgrass	L: 63–223 nm
Maize Stalk Leftover	L: 150–450 nm, D: 4–7 nm

Table 2.
Cellulose nanocrystals – Their sources and dimensions [7].

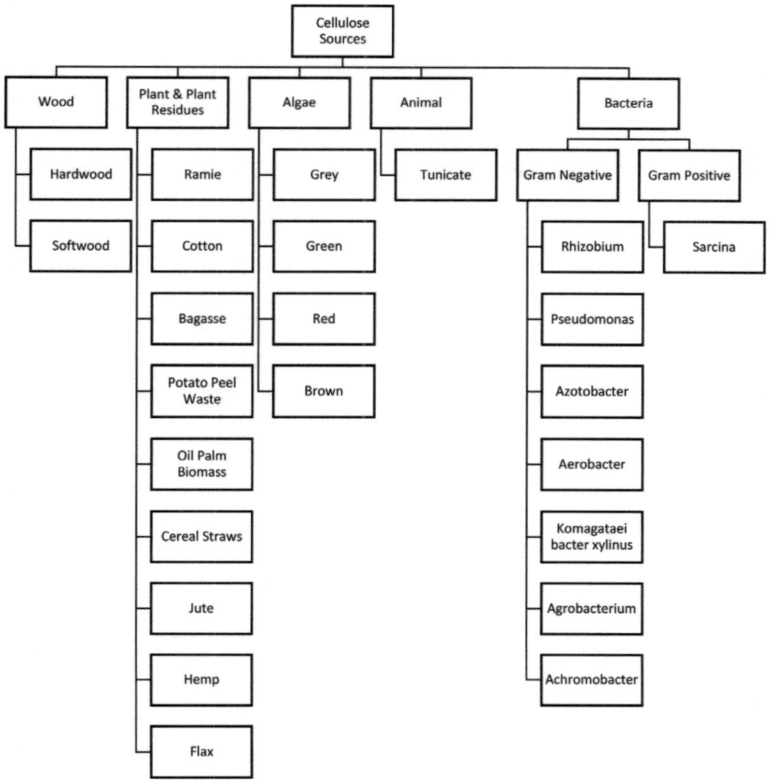

Figure 7.
Classification of cellulose according to sources [2].

higher surface area, and higher number of hydroxyl groups that can be easily accessed for the alteration of surface [14].

In comparison to CNCs, the fabrication of cellulose nanofibers (CNFs) is much easier because it does not need extreme chemical splitting to alter the molecular level structure of the cellulose chain. Normally, the fabrication of carbon nanofibers (CNFs) involves a wide range of physical or chemical techniques (**Figure 7**) [17].

3. Applications of nanocellulose

Nanocellulose has gained industrial and academia interest in the wide range of applications such as wettability, carrier of antimicrobial substances, barrier (i.e., UV protector and chemical and solvent protector), electrical, polymer reinforcement, biomedical, energy, automobiles, and smart and responsive materials are few of them.

3.1 Wettable application of nano-cellulose crystals

Wettability of substrate or surface refers to the investigation of how the liquid contacts on the solid substrate or surface, that is, it is balance between adhesion and cohesion forces. The angle of contact developed due to these forces between the liquid and substrate is called degree of wettability. The angle of contact (CA) subsequently evaluates to determine the hydrophilic, hydrophobic or hybrid nature of material or substrate [18, 19]. Because of the growing need in both the academic and industrial sectors, several scholars have been interested in fabricating hydrophobic nanomaterials in recent years. The hydrophobicity of nanocomposites is a crucial element in establishing their applicability. For industrial applications, there is a significant demand for hydrophobic materials that provide features such as self-cleaning capacity, antifouling properties, repellent to water, and lessened friction. CNCs are widely recognized for their exceptional strength and are commonly utilized as reinforcement materials. However, due to its strong hydrophilicity, CNC's effectiveness is presently inadequate as it could not be easily mixed into several matrices of polymer which are normally hydrophobic. Hence, hydrophobic alteration of CNC can result in enhanced dispersion in hydrophobic and nonpolar matrices. Additionally, surfaces of CNC with hydrophobicity may also be utilized as coating layer substance for maritime vehicles, medicinal devices, windows, fabrics, paints, and a variety of applications. The development of hydrophobic surfaces can open several ways for commercial usages of biopolymer which exist in nature [20].

Leaves of lotus are recognized for their ability to self-clean, have a rough hierarchical structure with two different degrees of roughness [21]. Numerous synthetic hydrophobic materials have been evolved due to the inspiration of superhydrophobic qualities of the lotus. The water contact angle (WCA) is a fundamental element to examine the hydrophobicity. A substance is hydrophilic if its water contact angle is $<90°$, if contact angle of water $>90°$ than substances is said to be hydrophobic, and superhydrophobic if its WCA is $>150°$ [22, 23]. To calculate WCA, a number of relations have been devised, including Wenzel equation, Young's equation, and Cassie equation [23]. The sliding angle is another crucial component of hydrophobicity along with the water contact angle. The sliding angle is the incline angle between a droplet and its substrate where the droplet begins to roll on the surface. The water repellent property of superhydrophobic material is frequently described as having a sliding angle smaller than 10° [23, 24].

Surface roughness or chemical modification can be utilized to control hydrophobicity. Surface energy drop through chemical alterations and higher roughness should be controlled simultaneously to achieve super hydrophobicity [25]. Materials with super hydrophobicity are produced using hazardous physical and chemical methods. Bonding of molecules with low surface energy, which includes fluorinated agents [26], silanes [25], organic hydrophobic chains [27], and phosphonates [28], and others are utilized in chemical modifications for the achievement hydrophobicity. **Table 3** shows a variety of chemical modifications and bonding to cellulose-derived products like cotton, paper, and nanocrystals of cellulose.

Though the chemical treatment has conventionally been employed to functionalize material surfaces for various applications, there is a recent inclination towards increasing of surface roughness, which has shown intriguing hydrophobic properties in several materials. Enhancing surface roughness is a critical factor in the improvement of water repellency [29]. Air, which is specifically hydrophobic in nature (WCA = 180°), becomes stuck in the roughness grooves [30] When a water droplet is in contact with the surface, it interacts with the trapped hydrophobic air, resulting in an increase in hydrophobicity [30]. Various surface roughness approaches, including etching, laser treatment, and electrospinning, are frequently employed [23].

Salajkova et al. [28] utilized quaternary ammonium salts alteration to produce hydrophobic alteration of CNCs. In this investigation, For the CNC alterations, four distinct quaternary ammonium salts were utilized. **Figure 8** depicts the addition of stearyltrimethylammonium chloride with three structure quaternary ammonium salts (1) glycidyl trimethylammonium chloride, (2) phenyltrimethylammonium chloride, (3) and diallyldimethylammonium chloride [28]. The maximum WCA of stearyltrimethylammonium chloride altered CNC was 71.0°C, though the notable advancements in WCA of the CNC surface was noticed, a greater WCA (>90°) is generally required for utilizing the CNC hydrophobicity for advanced material applications.

3.2 CNCs as carrier of antimicrobial substances

Since the most antimicrobial substances are tiny, therefore, there is always a chance that they will leach out of the material in which they are contained (such as fabric, and plastics, etc.). When the antimicrobial elements are lost to the garments dermis, and surrounding environment, the material becomes contaminated and causes health risk to human and the environment [20]. Due to the process of leaching

S#	Surface	Modification	WCA	References
1	Paper	TEOS and tridecafluorooctyl triethoxysilane	170°	[22]
2	cotton	Silica sol treated with PFSC	145°	[27]
3	CNC	Castor Oil	97°	[25]
4	CNC	Pentafluorobenzoyl chloride	112°	[24]
5	CNC	Stearyltrimethylammonium chloride	71°	[26]

PFSC: perfluorooctylated quaternary ammonium silane coupling agent.
TEOS: Tetraethyl orthosilicate.

Table 3.
Hydrophobic treatment of cellulose-based materials.

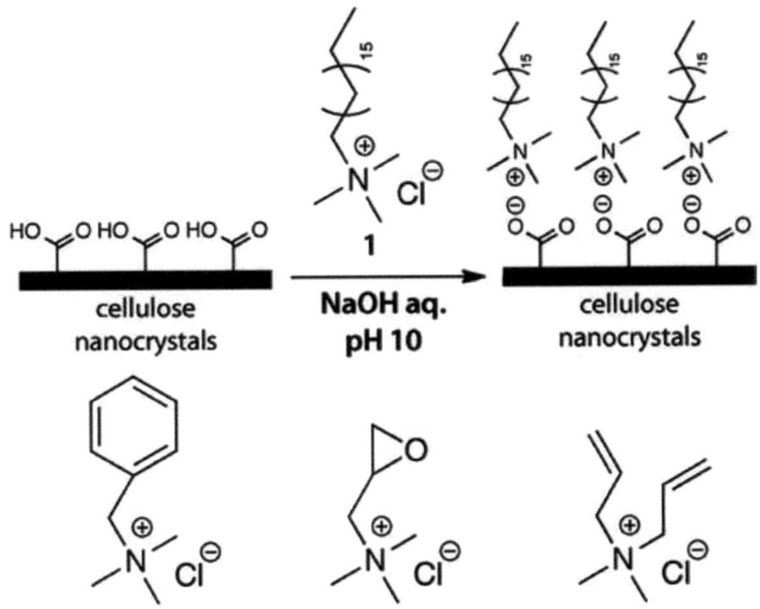

Figure 8.
CNC altered through y quaternary ammonium salts [28].

and the occurrence of undesirable interactions with substances of food, that is, fats and proteins, the immediate entry of antimicrobial substance leads to a reduction of its ability and efficacy [31]. A variety of antimicrobial substances including triclosan, iodine complex, phenol, and triclocarban, were detected in commercially available everyday items like soap, these antimicrobial substances were banned because they caused a human health risk [32]. Another driver for creative approaches to the development of antimicrobial carriers is the occurrence of resistance to antimicrobial drugs due to alteration of microbial and a declining efficacy of antimicrobial drugs in the treatment of usual microbial diseases [33]. There has been an increase in interest to convert antibacterial substances into chains of polysaccharides and polymers, particularly CNCs halogens, phenols, nanoparticles of silver, and salts of quaternary ammonium are frequently utilized as antibacterial substances. Antimicrobial characteristics also exist in transition oxides of metal oxides which include silver (Ag), gold (Au), magnesium (Mg), nitrogen monoxide (NO), titanium (Ti), copper oxide (CuO), iron (II, III) oxide (Fe_3O_4), and zinc oxide (ZnO) [34]. Antimicrobic peptides including magainins, cecropins, and defensins as well as antimicrobial enzymes like lactoperoxidase and lactoferrin are frequently utilized [35]. Quaternary ammonia compounds (QACs) with positive attach with a bacterium of negative charge. Once they diffused through the cell wall, they attack the surface cytoplasmic membrane, microorganisms struggle to survive and ultimately die as a result of the loss of vital cytoplasmic components [36]. So, QAC is one of the most active antimicrobial substances whereas, their poisonous effects on pathogens (fungi and protozoans), bacteria and mammals restrict their usage.

Nanoparticles of silver are recommended as antimicrobial substances because of their wide range of antimicrobial properties against a variety of pathogens, yeast, fungus, viruses, and both gram-negative and gram-positive viruses can all be killed off by silver nanoparticles [37]. The breathing function of the bacteria is disrupted and restricted via the silver particles. Nano sized silver particles are typically chosen because silver particles efficiency rises with the decreasing their size owing to an increase in surface area [38]. *E. Coli* (*Escherichia coli*) and *S. aureus* (*Staphylococcus aureus*) are both inhibited by silver particles in their growth [39]. Unwanted health effects of silver nanoparticles have restricted their usage for packaging of food in certain countries [37]. These health effects create limitations in research and application, specifically in the industries like food and medicine.

Triclosan is an artificial bio-phenol with high bactericidal activity (2,4,4′-trichloro-2′-hydroxydiphenyl ether). It kills both gram-negative and gram-positive bacteria, fungus, and mold [40], such, triclosan is used in variety of products, including mouthwash, soap, packaging, garments, and cooking kits [41]. The EPA (Environmental Protection Agency) regulates the triclosan usage because of its apparent poisonousness and ecological hazards, and due to that, its usage in everyday items is regulated [40].

Chitosan, a naturally occurring antimicrobial substance that is produced from the protein chitin, has become quite popular in industrial uses. Chitin is derived from exoskeletons of fungus, algae, and bugs [56]. It is an antibacterial and antifungal polycationic cellulose [42]. It also modifies or produces a polymers membrane on the cell surface, preventing nutrients from taking place and ultimately causing cell death [43]. Chitosan has been studied for a variety of uses including packaging, medication distribution, and number of biomedical applications [43]. But its previously mentioned activity demonstrated that it is an antimicrobial substance which works slowly, that makes it less potent than other antimicrobial substances. The most common application of antibacterial materials is in the food packaging industry. In addition to physical and moisture protection, successful packaging reduces the growth of microorganisms in the food. As a result, popularity of effective packaging is progressively raised to prevent, decrease, and eliminate microorganisms on the surface of food or in the vicinity next to the package. Recently, the demand for low preservative food from consumers has enhanced. Therefore, active packaging can be highly appealing for maintaining food quality under low preservation conditions. Active food wrapping film can extend shelf life of food product, preserve nutrient content, and offer microbiological safety while inhibiting pathogenic growth [44].

As the need of packaging with bioactive and biodegradable material rises, studies were carried out to develop effective films for packaging using Nisin as an antibacterial agent, as shown in **Figure 9**. Nisin, a 34-amino acid long bacteriocin, is efficient against a variety of gram-positive microorganisms found in food [45]. The utilization of nisin have been employed to stimulate chitin in order to produce an effective packaging material for pasteurized milk [46]. Furthermore, it was claimed that growth of micro bacteria on ground beef and oyster was slowed when film of activated nisin was utilized for packaging [46]. Nisin was integrated with polylactic acid-cellulose nanocrystal (PLA-CNC) compound [73]. In general, the PLA-CNC film with activated nisin exhibited all three fundamental characteristics of packaging materials, such as mechanical strength, antibacterial efficacy, and biodegradability. Weishaupt et al. [47] demonstrated the self-assembly bio composite of nanofibrillar cellulose-nisin. TEMPO-oxidized nano-fibrillated cellulose with carboxylic clusters produced a negative surface for nisin to be adsorbed onto nisin binding to nanocellulose has significant impact through electrostatic interactions. Meanwhile, the binding ability of nisin and nanocellulose was reduced at elevated concentrations of salt [48]. Nanocellulose bounded with nisin was tested against the *S. aureus* and decline in growth was found [48]. In contrast to traditional antimicrobial carrier

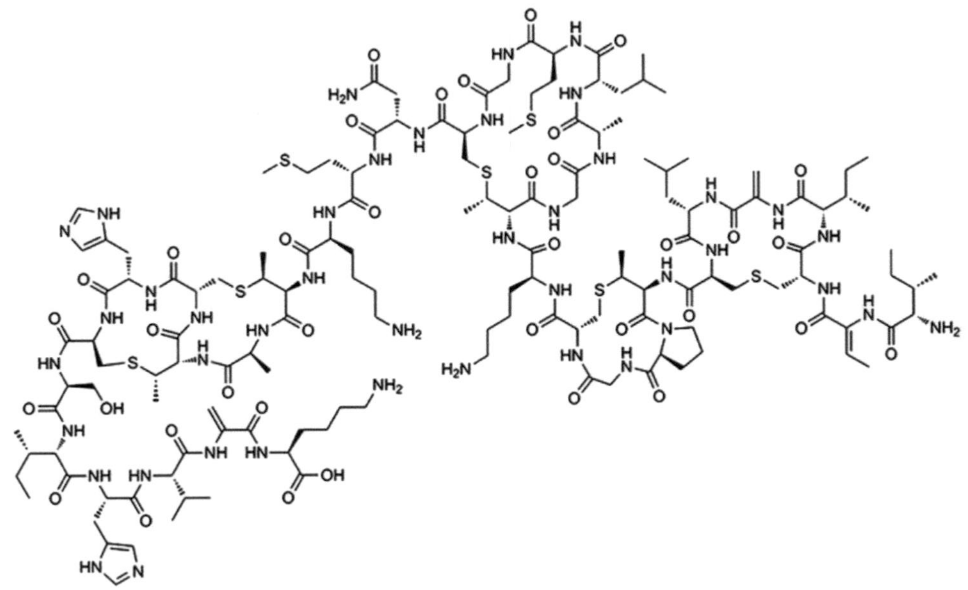

Figure 9.
Nisin chemical structure. Replicated with approval from Salmieri et al. [49].

agents, nanofibers can additionally offer mechanical strength, that is, essential for many applications. Which is presently an effective research field, with an emphasis on getting improved antibacterial characteristics, and optimizing cost, scale-up this research in order to commercialize.

3.3 Applications of CNCs as barrier

Restricting the environmental impact on materials has always been a prominent priority in many sectors. Corrosion is an ongoing issue and concern in oil and gas industries, where a pipping network is the main transporting system for delivering gas and oils. The interaction of transporting system to sea water in the marine industry is continually corroding the system. Interaction of polymers with UV lights over time can affect its characteristics at micromolecular level. Except for a few academic experiments, the usage of CNCs as an application of barrier is very less [20].

3.3.1 UV protection

The visual appearance of polyurethane (PU) begins to alter gradually when exposed to UV light, causes the yellowing of the material due to photochemical degradation on the molecules surface. Zhang et al. [50] conducted a study to employ CNCs as UV stabilizer for the prevention of photochemical degradation. 3-Glycidyloxypropyl trimethoxy silane (GPTMS) was utilized to alter the CNC at numerous concentrations. After the GPTMS hydrolyzing, the CNC was added and allowed to react throughout a specific duration. **Figure 10(a)** depicts the suggested salinization mechanism of reaction and interaction among the altered PU and CNC. The altered CNC was subsequently mixed in the PU mixture and homogenized to disperse the altered CNC in the PU as presented in **Figure 10(b)** to establish the

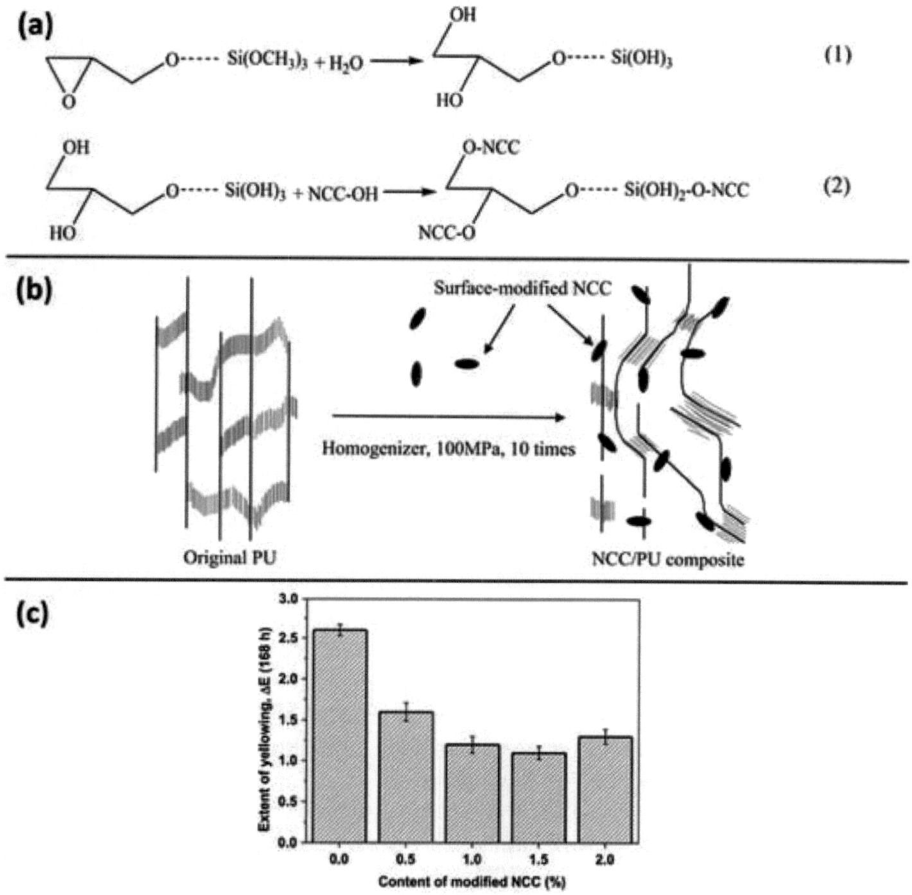

Figure 10.
(a) Shows a CNC alteration reaction scheme for the utilizing 3-Glycidyloxypropyl trimethoxysilane (b) shows a graphic illustration of altered CNC dispersion inside PU. (c) Shows a bar chart indicating the impact of altered CNC amount on the PU yellowing being exposed to UV light throughout an established time period [50].

composite of CNC & PU. This mixture gradually opens to UV light. The outcomes of investigation revealed that inclusion of an altered CNC in the PU significantly decreased the UV light yellowing impacts, further reduction in UV light yellowing impacts was noticed with the raise of alter CNC concentration. With the incorporation of 1.5% of altered CNC, the yellowing impact was decreased around 58.0%, demonstrating the usefulness of CNC as an antiyellowing substance as shown in **Figure 10(c)**. It was hypothesized that altered CNC reduced photo degradation of the CH_2 cluster while prohibiting scrubbing of urethane cluster.

In other research, CNC was utilized as dual-purpose filler that gives polyvinyl alcohol polymer reinforcement and a UV barrier [80]. Pulp of CNC was oxidized through sodium metaperiodate and then combined in HCl solution with sodium 4-amino-benzoate to create altered CNC that had been affixed to photoactive groups. This modified CNC was then mechanically decomposed to create p-aminobenzoic acid grafted CNC (PABA-CNC). Numerous concentrations of this altered CNC were introduced to a polyvinyl alcohol (PVA) water solution. The resulting mixture solution was subsequently degassed and cast to form thin coating, that were utilized for additional testing. The UV transmission outcomes with the adding of PABA-CNC to

PVA demonstrated that the existence of PVA with PABA-CNC drastically decreased UV transmittance in contrast to alone PVA. The rising concentration of demonstrate the additional decline in UV transmission with PVA coatings comprising of 0.5 and 10% PABA-CNC decline the transmission 54.0 and 12.0% respectively, as compared to clean PVA coating, which indicated 70.0% transparency. PABA-CNC inclusion also improved the mechanical characteristics of thin coating, such as tensile strength and modulus, this rise was a function of PABA-CNC concentration. Ethyl cellulose nanoparticles (ECNPs) have also been investigated as a method of confining or protecting UV light, that tends to produce species of oxygen that are reactive when subjected to UV because of photodegradation [50]. Such filters are typically employed in makeup, such as sunscreens, and can absorb cancer causing reactive species of oxygen, that could be harmful when it comes into skin contact [20].

3.3.2 Nanocellulose as protector of corrosion and chemicals

Surface coatings are commonly used in numerous industries to protect their products and equipment's surface from surrounding environment. Some of the usage involve the application of paint to retard the rust on metallic substances, the employment of transparent epoxies on wood and plastic to avoid the scratching, and to avoid the effects of ultraviolet (UV) deterioration and decomposition. However, transparent coat epoxy resins or paints for corrosion protection do not always possess adequate strength and resistance to chemicals. Ma et al. [51], conducted a study on the characteristics of CNCs, and reported its strengthening abilities at minimal loadings because of their crystallinity and nano-size, which contributed to their utilization in the application of metals coating as a base element in epoxy. In this investigation, epoxy including 1.00, 1.50, and 2.00 wt.% CNC was properly homogenized with a rod of glass and sonicated to make sure appropriate dispersion before being painted with brush in a very thin coats on mild steel. The coated steel was permitted to dry before it was evaluated for corrosive resistance utilizing electrochemical impedance spectroscopy (EIS) during a 30-day dip in 3.5% sodium chloride. Moreover the transparency of the coating epoxy was examined through UV-vis evaluation to detect the influence of CNC. The optical transmission findings demonstrated that as CNC dosage increased, the rate of transmission reduced to 20.0% for coating comprising 2.0 wt.%. In contrast, the coating comprising 1.0 wt.% CNC was found to be exceedingly clear, with transparency of 74.0%.

Additionally, it was noticed that the light drop off transmittance for all nanocomposite was between 300 and 350 nm, indicating significant light absorbance and the absence of light reflections in the UV region between 300 and 400 nm. It demonstrates that this type of coating can be utilized to stop UV deterioration but also retains outstanding visibility when utilized for transparent coating applications, particularly at 1 wt.% of CNC. While the corrosion evaluation revealed that the application of epoxy coating significantly retards the corrosion by the addition of CNCs. This might be as a result of the CNC acting as an obstacle to the sodium hydroxide ions solution by directing them into a tortuous path. This prevented the coating's contact entirely, therefore shielding the surface of the mild steel. After 1 day of being exposed, only the unreinforced epoxy experienced penetration. Neat clean epoxy coating exhibited two-time constants when the coatings were electrochemically analyzed using bodes plots. On the 30th day, only the unaltered epoxy kept producing two-time constants, demonstrating the CNC's outstanding anticorrosion properties. But as the test went on over the course of the 30-day period, for the epoxy coating with a 2 wt.% CNC loaded, the presence of a two-time constant gradually emerged. It was ascribed to the CNCs potential for aggregation within the epoxy, as they did not scatter uniformly in 2 wt.% like they did in the samples loaded at 1 and 1.5 wt.%. This irregular scattering lead to corrosion of the mild steel occurred from the ions (Na^+ and Cl) which being able to diffuse into the epoxy's exposed areas and cause corrosion [45].

3.4 Nanocellulose in electrical and electronic applications

Study in the development of functional CNC applications in electrically active materials, such as dielectric and electric conductive materials have great future potential, whereas materials like starch [24], proteins and peptides [52] currently being investigated for the same purposes. The popularity of CNCs in electrical industries is raised due to their ease of manipulation, piezoelectric and dielectric capabilities, and sustainable characteristics similar to other bioderived materials. Csoka et al. [53] investigated extremely thin films comprising highly oriented c through a tip of diamond throughout extremely thin films ellulose nanocrystals which demonstrated piezoelectric impacts due to the cumulative production of the separate CNCs. Extremely thin films with multiple levels of CNC orientation were created using a method is reported in another paper [54]. The atomic force microscopy (AFM) in tapping mode was utilized for the displacement measurement of the film when an electric field was employed to them. Higher piezoelectric effect was noticed with a higher degree of alignment. Thus, the film's piezoelectricity was based upon the CNCs alignment. The determined results are not merely attributed to the CNCs alignment but also owing to the CNCs crystallization in films and orientation. **Figure 11(a)** exhibits the AFM diagram in tapping method comprising aligned CNCs by employing an electric field across it whereas **Figure 11(b)** illustrates the applied voltage impact on thin films due to the CNCs displacement. The outcomes of this investigation reveal that extremely thin films by numerous levels of CNC alignment can end up to various degrees of electro-mechanical actuation, that may possibly be employed in different applications, for example, extremely sensitive weighing scales. Also the possession of a high-level CNC concentration and positioning might lead to the application of very delicate forces.

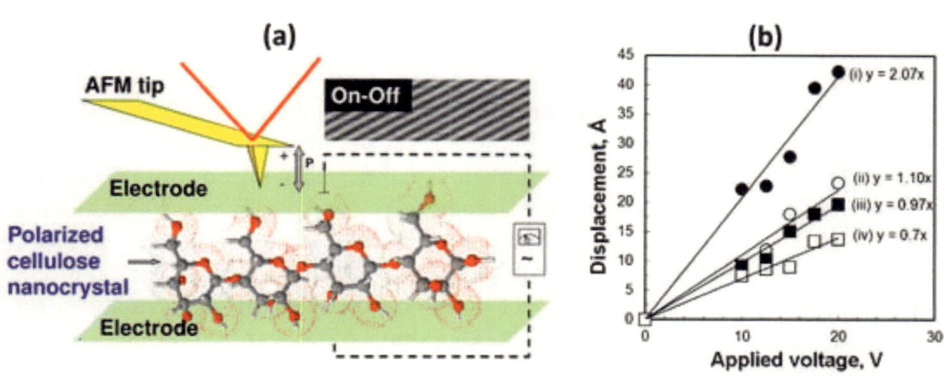

Figure 11.
(a) Dielectric characterization of ultrathin film coated with properly aligned CNC via atomic force microscopy (AFM) in tapping approach and (b) voltage impact on displacement of CNC (piezoelectric effect) [53].

CNCs can alter the polarization densities because of the greater level of crystallization. Similarly, its dielectric character enables its utilization as an effective insulating material in various purposes. The existence of moisture in the CNC plays an essential role in identifying the dielectric characteristic, because it behaves like an electrical conductor when moist exists in CNC. For the utilization of CNC in applications of dielectric, the levels of moisture should normally be around 0.50% [55], because of its hygroscopicity, it normally has a level of moisture in the range of 4.0–8.0%. These moist contents range stems from CNCs source and extremely dependent on the cellulose crystallization as studied from cellulose water sorption investigations [56]. Overall, moisture content will be lowered with the higher cellulose crystallinity. Bras et al. [55] investigated the dielectric characteristics of two different wood nanocelluloses, nano-fibrillated and algal (cladophora cellulose) nanocelluloses [55]. But they also observed the better sorption potential of NFC at both high and low moisture owing to its lesser crystalline. Even though it was thought that dielectric qualities were closely associated with crystallinity, NFC was found to have a higher dielectric characteristic than cladophora cellulose. It was because of the high porosity of cladophora cellulose, that permits air to get inside and raise its dielectric loss. It is evident that nanocellulose have good dielectric characteristics that might be used to insulate electrical wires and cables, however, the efficacy of this feature relies not just on the source but also on the nanocellulose shape [57]. It additionally demonstrated that NFCs made from CNC might be utilized for the manufacture of wood, glass and polymer dielectric substances [58].

3.5 CNC application in automobiles

Multi-functional nanocomposites not only exhibit enhanced mechanical strength, but also demonstrate electrical, optical, thermal, and magnetic characteristics. Molecular level interaction among the polymer matrix and nanomaterials, as well as the presence of interface region among the bigger polymer-nanomaterial, are assumed to play crucial part in effecting the physical and mechanical characteristics. CNC has been gaining lot of attention due to its increased mechanical characteristics, constancy in dimension and raised in modulus, flammable obstruction, outstanding thermal and process properties, and improved resistance effect, creating it more demanding substitute of metal in applications of automobiles [59, 60]. The fundamental purpose of utilization of nano-polymer in automobiles industries is to reduce weight of vehicles, enhanced efficiency of engine, CO_2 emissions reduction and enhanced performance. The polymeric composites commercial utilization starts from 1991, at that time Toyota Motor Co. introduced bio composites of nylon 6/clay in market to produce belt. During the same time, Unitika Co. of Japan launched engine cover developed from nanocomposite of nylon 6, which offered highly finished surface and weight reduction by 20%. General Motors in different phase introduced GM safari parts manufactured from polyolefin with coating of 3% nano clay. In 2002 Chevrolet Astro vans, in partnership of Basell (now LyondellBasell Industries), used nanocomposites of polymers in doors of Chevrolet Impalas [61]. During 2009 General Motor manufactured rear-floor in single piece through compression molding assembly for Solace Pontia from nano-enhanced sheet. This single piece rear floor is also utilized for GM's Corvette ZO and Chevrolet Corvette Coupe. Automobiles industry can take advantage from nano-polymer in wide range of application for example braking systems and suspensions, body component and frames, power systems and engines, catalyst converters and systems of exhaust, paints, tires, lubricants, and electronic and electrical component etc. A scientist group of Japan Toyota Central Research Development Laboratories recognized work on nanocomposite of nylon-6/clay in late 1980s and in early 1990s developed upgraded procedures for the manufacturing of nanocomposites of nylon-6/clay using in situ polymerization alike to the Unitika procedure. The research outcomes demonstrated a key development in mechanical and physical characteristics via reinforcement of nano-polymers with clay. The same team also documented similar methods for several different polymers-based nanocomposites clay forms, that is, acrylics, epoxy resin, polystyrene, polyimides and elastomers. After that detailed research has been conducted throughout the world in the field of nanocomposites [62, 63]. The nano-polymers tiers with greater scratch resistance and catalysts for fuel-borne raises inhibitions of soot penetrability in specific filters, windshields and headlights anti-fog coatings and scratch resistance coating of vehicles body etc. The advancement in the nanocomposite R&D developments is continuously increasing due to the interest of financial agencies and organizations, as these organizations want to expand the capitalization on newly discovered products. Such as, European automobile industry spends nearly 5% of annual revenue on R & D. The primary aim is to produce effective and stronger coatings and paints. Therefore, utilization of nanocellulose composite polymer in automobiles industries is projected to raise in coming years [64].

3.6 CNCs as an applications of other advanced functional materials

CNCs are employed for wide range of applications owing to the hydroxyl clusters on their surfaces which can be functionalized to do various activities. CNCs produced by hydrolysis of sulfuric acid have negative charge on the surface, which makes electro statistical repulsion among the CNCs. Due to this, CNC is simply dispersed in polar polymer frameworks like PVA. In the past 20 years, both academic and business experts have been interested in CNC utilization. Apart from that it can be used as a reinforcing substance, CNCs have shown promising usage in biomedical science, personal care, and energy.

3.6.1 Polymeric reinforcement

Nanocrystals of cellulose are frequently utilized for filler reinforcing in matrices of polymer to raise their strength. Conventional polymers did not have their desired strength for the majority of structural applications, so they need fillers to strengthen them. Being a biodegradable material, CNC offers an outstanding polymers reinforcement. Numerous investigations have demonstrated that very low quantities of CNC enhanced the physical and thermal characteristics of polymers because of their nano-size and capability to effectively absorb matrix strength [65]. Bras et al. [66] stated the CNC impact on reinforcing the rubber material. They observed that CNCs could enhance both mechanical and thermomechanical characteristics of rubber. But, because of the hydrophilic characteristic of CNCs, rubber absorbed more water. This investigation demonstrates that CNCs could be utilized to improve the properties of polymers in certain applications where the composites will not be exposed to moisture. Moreover, Cao et al. [67] demonstrated the reinforcing abilities of CNC in cement. It was suggested that adding around 0.2 vol.% CNC raised the hydration of the paste, resulting in a 30% rise in flexural strength. These examples demonstrate that the nano-polymer could be the composite materials for future utilized in automobiles parts, furniture, construction, and several high strength materials.

CNC was effectively attached with several kinds of polymers to alter its surface and acquire the altered CNC with the desired characteristics. For example, the utilize of hydrophobic polymer attached with non-polar polymers commonly enhanced the interaction and dispersion among the CNC and polymers as well as raises the composite polymers strength. In some other situations, the attached polymer increases functionality and widens the applicability of CNC. **Table 4** illustrates numerous polymers attached to CNC and accomplished the characteristic owing to the alteration.

3.6.2 Biomedical applications

CNC application in biomedical involves its utilization in making the medical devices, wound recovering, bioimaging, tissue engineering structure, and delivery of controlled drugs [68]. Dong and Roman observed the bioimaging application with fluorescent labeled CNC [69]. In those investigations, epichlorohydrin was utilized

S#	Polymer	Reaction Mechanism	Conclusion	Reference
1.	POLY(4-VINYLPYRIDINE) (PVP)	Radial polymerization with ceric ion	Thermal degradation enhanced by 60°C temperature	[66]
2.	POLYSTYRENE (PS)	Atomic Transfer Radical Polymerization (ATRP)	Enhanced the water contact angle by 50%. 1,2,4-Trichlorobenzene (1,2,4-TCB (organic Pollutant) can absorb 50%	[67]
3.	POLYMETHYL ACRYLATE (PMA)	Polymerization with mediation of nitroxide	CNC attached PMA was around 1% weight. Was soluble in acetone and not dispersible in water.	[68]
4.	POLY TERT-BUTYL ACRYLATE (PTBA)	Grafting to	Stable suspension in acetone, chloroform, and toluene	[69]
5.	POLYLACTIC ACID (PLA)	Ring Opening Polymerization (ROP)	Well dispersed in chloroform (non-polar solvent)	[20]
6.	JEFFAMINES (POLYETHER AMINE)	Grafting to	Thermo-reversible aggregation CNC grafted Polymer has lower surface charge.	[70]
7.	POLYMETHYL METHACRYLATE (PMMA)	Radical polymerization atomic transfer	Reduced thermal stability (250-290°C). Enhanced water contact angle by 17°.	[71]
8.	POLY(ε-CAPROLACTONE) (PCL)	Ring Opening Polymerization (ROP)	Addition of CNC attached PLC matrix enhanced young's modulus from 231 to 582 GPa. For the same matrix elongation break declined significantly.	[72]
9.	POLYETHYLENE GLYCOL (PEG)	Grafting to	Thermal decay of CNC graft appears at lower temperature.	[73]

Table 4.
Nanocellulose grafted with polymer.

for binding fluorescein-5′-isothiocyanate to CNC. To study the biodistribution and interactions of CNC in living organisms using fluorescence labelling, that might be meaningful for several biomedical applications. This investigation demonstrated that the enhanced CNC might be utilized to analyze the cells interaction. The researchers additionally stated that the functionalized CNC was being utilized in a different study to look at how it interacts with cells from mammals [20]. Drug transporters can occasionally trigger immunological reactions in systems associated with biology, for example the human body. An investigation [20] regarding the application of cyclodextrin carriers CNC demonstrated that it was effective in drug delivery application and had no negative side effects. Immunological responses, for example, the transmission of the pro-inflammatory cytokine interleukin 1 (IL-1), and mitochondria-derived reactive oxygen species (ROS), were found to be negligible [70]. CNC is deeply investigated for tissue engineering, in which supporting equipment was utilized for self-recovering and regrowing. In order to enhance performance, selection of proper material for scaffolding is very important. Physical, biological, and mechanical characteristics are playing crucial role in high efficiency and effective mechanical combination. Therefore, characteristics like porosity, size of pore, shape, area of surface to volume ratio, and roughness of the surface should be considered. Furthermore, the rate at which organic substances degrade plays an essential role in healthy tissue healing whereas the scaffolding material gets absorbed [20]. In most situations, unfilled polymer materials cannot be utilized to attain both biological and mechanical characteristics. Hence, nanofillers are used to create nanocomposite materials with functional characteristics including electrical conductivity, self-assembly ability, mechanical strength, and adhesiveness [71]. Domingues et al. [71] comprehensively specified the utilization of CNC-PLA composite for tissue engineering to satisfy the previous parameters for biomedical scaffold and its use in their investigation on CNC-based biological material for tissue engineering [72].

3.6.3 CNC applications in energy

CNC application in energy implies the utilization of composites materials produced from cellulose for storage of energy. With the rising of environmental and sustainability concerns in making more effective and feasible resources of renewable energy. Zhou et al. [73] designed and developed recyclable solar cell from cellulose nanocrystal by utilizing the benefits of outstanding mechanical characteristics of silver and CNC [20]. A semitransparent, recycling solar cell with electrodes was fabricated. But more improvements were required to meet the essential performance and efficiency. Kim et al. [74] stated some other energy applications of cellulose based materials like display devices, energy harvesting machines, paper transistors, and motors, where the CNC with outstanding biocompatible, mechanical characteristics and easiness in functionalization process provides its capacity to present sustainable and ecofriendly technologies.

3.6.4 CNC as a responsive and smart materials

In recent years, the responsive and smart materials utilization has risen. The responsive and smart materials modify the external environment and deliver responses. The alteration in impetus like acquaintance to heat, chemicals, light and magnetic fields might be utilized to produce mechanically adaptable materials that reply to stimuli nanocomposite and could react towards outside stimuli in several ways, for example by expanding or shrinking, and assemble and dissemble etc. [75]. These alterations can be utilized as a stimulus responsive smart material. CNC could be utilized as a material that responds to different stimuli for sensors and other products. It can adapt to respond to light, pH, heat, moisture, chemicals, and magnetic fields, which offers to respond the inputs along with the ability to reinforce. When the pH changes, the rheological characteristics of CNC composites also alter [76]. Way et al. [76] developed carboxylate and amine-functionalized CNC in order to evaluate the response of pH. Also, by modifying the CNC surface chemistry, the nanocomposites could be modified to produce numerous materials for mechanical adoption.

Smart sensors based on CNC could be designed to detect ions, moisture, biological species and organic vapors. Kafy et al. [77] developed a humidity sensor from a CNC-graphene oxide (GO) composite. CNC-GO coating demonstrated enhanced absorption of water, that is beneficial for moist sensitivity [77]. The sensing film did not compromise on its performance with the variation of temperature, illustrating the practical usage of a moist sensor [77]. CNC might additionally be customized to develop sensing material for gas, capable of detecting different organic and hazardous gases. Furthermore, sensing material based on CNC can be utilized to recognize ionic substances. For the detection of ferric (Fe^{3+}) ions, CNC containing pyrene was produced [78]. This idea could be more extended to develop a material to sense the dissimilar ions, biological and chemical compounds. Some other smart sensors developed from CNC comprised of strain and proximity sensor, which was investigated by Sadasivuni et al. [79] and Wang et al. [80].

4. Economic and environmental evaluation of nanocelluloses

4.1 Economic evaluation of nanocelluloses

Manufacturing cost of nanocellulose depends on the methods of pretreatment, acid utilized in hydrolysis process, and cost of feed stock. Nanocellulose manufacturing cost stated in previous study ranges from 2000 to 2010,000 USD per ton [81, 82]. The manufacturing of CNC through sulfuric acid hydrolysis requires acid-resilient apparatus that raises the capital expenses (CAPEX). Furthermore, reclamation of acid through waste treatment and neutralization increases capital costs. Research on CNCs stated that sulfuric acid hydrolysis production costs varied from 3632 to 4420 USD per dry ton of CNCs [81]. In all the cases pulp dissolving cost was the primary contributor, which contributes around 38–45% of production cost. For zero acid reclamation, lime and acid costs contribute around 24–38% of production cost. While for acid reclamation, this contribution considerably reduced to 3–4% of production cost. The CNC minimum selling price (MSP) was mainly affected by the cost needed for pulp dissolving and CAPEX. The raise in CAPEX because of the acid reclamation, increased the MSP from 4829 to 5125 USD per CNC dry ton [81]. A study conducted by Blair et al., [83] stated that cost of equipment contributes 65% of CAPEX. Another investigation stated that CNCs acquired through citric acid hydrolysis are more environmental friendly than sulfuric acid hydrolysis, but MSP was comparatively greater and was 16,460 USD per dry ton [84]. The utilization of citric acid hydrolysis in CNC manufacturing shows a 7 million USD reduction in capital investment than sulfuric acid hydrolysis. However, the operational expenditures rose to 8.25 million USD per year, where citric acid alone accounting for 40% of OPEX at 89% acid reclamation. It means that CNC manufacture using citric acid hydrolysis will be more expensive than sulfuric acid hydrolysis. Presently, nanocelluloses are not economically viable when compared with alternate options because of their small manufacturing capacity. Shen et al. [85] evaluated the variables that impede the commercialization of nanocellulose reinforced polymer composites. The researchers stated that the pretreatment expenses and small manufacturing capabilities related to nanocelluloses lead to considerably increase the production in comparison to alternative options such nano montmorillonite and nano silica. In this regard, process optimization is crucial to attain mass production for the commercialization of nanocelluloses. Furthermore, the absence of a standardized product index, because of the variable characteristics of nanocelluloses such as synthesized from a wide range of sources and manufacturing methods, obstructs the commercialization of the nanocelluloses. Enzyme-mediated manufacturing of nanocelluloses provides a promising option to successfully tackle both environmental and economic challenges. Nevertheless, it is important to acknowledge that commercial enzymes are frequently utilized to efficiently convert cellulose and hemicellulose components into fermentable sugars. Additional investigation is required to enhance the efficiency of enzyme mixtures for the creation of nanocellulose [86]. Furthermore, the economic feasibility can be improved by producing enzymes on-site [87]. Previous investigations on nanocellulose manufacturing predominantly utilized bleached chemical pulp or pure cellulose, which had already lost biomass content by 50–60% for the manufacturing of nanocellulose [88]. This approach led to less effective employment of biomass because of the low nanocellulose yields in comparison to the original biomass. In contrast, mechanical pulping can serve as an effective substitute and requires more investigation. Additionally, research efforts must be directed towards optimizing the reaction yield to enhance the total economics of the manufacturing method [89]. Furthermore, technologies that are feedstock agnostic can play an important role in improving manufacturing scale. In such a scenario, economies of scale can be used to acquire cost-effective manufacturing. Future studies should be concentrated on approaches with moderate reactions and single-step extraction methods to enable the economical nanocellulose manufacturing.

4.2 Environmental evaluation of nanocelluloses

The growing utilization of nanocellulose in diverse industries and applications raises the probability of human contact with it throughout the different phases of the products. While nanocellulose is generally considered to be non-toxic, there are still uncertainties regarding its effects on the environment and human health. Furthermore, there are currently no established industry regulations or consumer product risk-assessment protocols in place to evaluate the health, safety, and environmental effects of nanocellulose before it is used in commercial products. This is because there is a lack of experimental data on the exposure of nanocellulose to both in vitro and in vivo systems under environmental conditions.

In order to quantify the environmental risk associated with nanocellulose, Natasha et al. [90] characterized both environmental exposure and hazard. The findings indicate a risk characterization ratio (RCR) of 6.9×10^{-5} for 2015 and 7.1×10^{-4} for 2025, indicating that based on the selected assumptions and a compound annual growth rate (CAGR) of 19% for nanocellulose production in the coming years, there is no current or anticipated environmental hazard associated with nanocellulose. Also, in accordance with the five-step procedure outlined in the framework, Piccinno et al. [91] conducted a scaled-up life cycle assessment (LCA) study for future nanocellulose production using a novel nanocellulose production pathway. They reported that the environmental impact per kilogram of manufactured nanocellulose yarn can be reduced by a factor of up to 6.5 when compared to laboratory production and suggested that the environmental impact of commercially available nanocellulose would be more comparable to that of an actual manufacturing facility.

The environmental impact of nano-fibrillated cellulose produced from thermo-groundwood, which involved the removal of extractives, lignin, and hemicelluloses, as well as TEMPO oxidation and homogenization procedures, was assessed using a Life Cycle Assessment by Turk et al. [92]. Their data indicates that the purifying procedure accounts for almost 95% of the overall impact, which is linked to a comparatively elevated usage of electrical energy and additional chemicals, specifically cyclohexane and acetone. They also reported that 1 kg of nano-fibrillated cellulose has a global warming potential similar to 800 kg of CO_2, and basic energy consumption is ~19 MJ/kg. The study also aimed to achieve a methodological objective, specifically by calculating the impact indicators using the three most significant assessment methods: ILCD/PEF, CML 2001, and ReCiPe 2016. These three strategies yield comparable outcomes in terms of their effects on global warming and acidification.

5. Conclusion

Cellulose is an organic polymer which has been utilized since hundreds of years as a fiber or its derived substances in multiple applications. Nanocellulose is an eco-friendly material and in advanced technologies its numerous applications are gradually becoming significant. The functionalization and flexibility of nanocellulose has the potential of application for the sustainable future. These materials are biodegradable and nontoxic for multiple applications, particularly considering the present environmental issues such as climate change caused by greenhouse gases released from the extraction and utilization of petroleum-derived materials. It has received special attention because of its availability throughout the world, replenish ability and the rapidly expanding technologies on its manufacturing. Thus, it is predicted that in the coming years, the nanocelluloses will become extensively and broadly utilized in automobiles, energy sector, medical, biotechnology, electrical and food industries etc. As nanocellulose continuously gains the interest of scientists all around the world, therefore enhanced understanding will expedite technology improvements.

Acknowledgements

This study is supported financially under the Fundamental Research Grant Scheme awarded by the Ministry of Higher Education Malaysia with grant number: FRGS/1/2019/TK03/UM/02/12 (FP143-2019A). The authors also gratefully acknowledge the support from grants, RMF0400-2021, ST049-2022, RK001-2022 and Department of Mechanical Engineering, Universiti Malaya to conduct this research work.

DOI: http://dx.doi.org/10.5772/intechopen.114221

Author details

Kaleemullah Shaikh[1,2], Wajahat Ahmed Khan[3], Md. Salim Newaz Kazi[1]*
and Mohd Nashrul Mohd Zubir[1]

1 Faculty of Engineering, Department of Mechanical Engineering, Universiti Malaya, Malaysia

2 Faculty of Engineering, Balochistan University of Information Technology, Engineering, and Management Sciences (BUITEMS), Quetta, Balochistan, Pakistan

3 Nanotechnology and Catalysis Research Centre (NANOCAT), Institute for Advanced Studies, Universiti Malaya, Malaysia

*Address all correspondence to: salimnewaz@um.edu.my

IntechOpen

© 2024 The Author(s). Licensee IntechOpen. This chapter is distributed under the terms of the Creative Commons Attribution License (http://creativecommons.org/licenses/by/3.0), which permits unrestricted use, distribution, and reproduction in any medium, provided the original work is properly cited.

References

[1] Phanthong P, Reubroycharoen P, Hao X, Xu G, Abudula A, Guan G. Nanocellulose: Extraction and application. Carbon Resources Conversion. 2018;**1**(1):32-43. DOI: 10.1016/j.crcon.2018.05.004

[2] Seddiqi H et al. Cellulose and its Derivatives: Towards Biomedical Applications. Vol. 28, 4. Netherlands: Springer; 2021. DOI: 10.1007/s10570-020-03674-w

[3] Dufresne A. Nanocellulose: From Nature to High Performance Tailored Materials. Berlin, Germany: Walter de Gruyter GmbH & Co KG; 2017

[4] Huang J, Dufresne A, Lin N. Nanocellulose: From Fundamentals to Advanced Materials. Lausanne, Switzerland: John Wiley & Sons; 2019

[5] Missoum K, Belgacem MN, Bras J. Nanofibrillated cellulose surface modification: A review. Materials. 2013;**6**(5):1745-1766. DOI: 10.3390/ma6051745

[6] Lee K-Y. Nanocellulose and Sustainability: Production, Properties, Applications, and Case Studies. Lausanne, Switzerland: CRC Press; 2018

[7] Zinge C, Kandasubramanian B. Nanocellulose based biodegradable polymers. European Polymer Journal. 2020;**133**(May):109758. DOI: 10.1016/j.eurpolymj.2020.109758

[8] Klemm D et al. Nanocellulose as a natural source for groundbreaking applications in materials science: Today's state. Materials Today. 2018;**21**(7):720-748. DOI: 10.1016/j.mattod.2018.02.001

[9] Heise K et al. Nanocellulose: Recent fundamental advances and emerging biological and biomimicking applications. Advanced Materials. 2021;**33**(3):e2004349. DOI: 10.1002/adma.202004349

[10] Randhawa A, Dutta SD, Ganguly K, Patil TV, Patel DK, Lim KT. A review of properties of nanocellulose, its synthesis, and potential in biomedical applications. Applied Sciences (Switzerland). 2022;**12**(14):7090. DOI: 10.3390/app12147090

[11] Thomas B et al. Nanocellulose, a versatile green platform: From biosources to materials and their applications. Chemical Reviews. 2018;**118**(24):11575-11625. DOI: 10.1021/acs.chemrev.7b00627

[12] Isogai A. Emerging nanocellulose technologies: Recent developments. Advanced Materials. 2021;**33**(28):1-10. DOI: 10.1002/adma.202000630

[13] Sapuan RAISM, Norrrahim MNF. Industrial Applications of Nanocellulose and its Nanocomposites. Cambridge, USA: Elsevier Science; 2022

[14] Hanieh Kargarzadeh AD, Ahmad I, Thomas S. Handbook of Nanocellulose and Cellulose Nanocomposites. Weinheim, Germany: John Wiley & Sons; 2017

[15] Mateo S, Peinado S, Morillas-Gutiérrez F, La Rubia MD, Moya AJ. Nanocellulose from agricultural wastes: Products and applications—A review. PRO. 2021;**9**(9):1594. DOI: 10.3390/pr9091594

[16] Pradhan D, Jaiswal AK, Jaiswal S. Emerging technologies for the production of nanocellulose from lignocellulosic biomass. Carbohydrate Polymers. 2021, 2022;**285**:119258. DOI: 10.1016/j.carbpol.2022.119258

[17] Ramesh Oraon CMH, Rawtani D, Singh P. Nanocellulose Materials Applications, Fabrication and Industrial. 1st ed. Cambridge, USA: Elsevier; 2022

[18] Shaikh K et al. Mitigation of $CaCO_3$ fouling on heat exchanger surface using green functionalized carbon nanotubes (GFCNT) coating. Thermal Science and Engineering Progress. 2023;**42**:101878. DOI: 10.1016/J.TSEP.2023.101878

[19] Shaikh K et al. Investigation of zirconium (Zr) coated heat exchanger surface for the enhancement of heat transfer and retardation of mineral fouling. Journal of the Taiwan Institute of Chemical Engineers. 2023;**153**:105246. DOI: 10.1016/j.jtice.2023.105246

[20] Panchal P, Ogunsona E, Mekonnen T. Trends in advanced functional material applications of nanocellulose. PRO. 2019;**7**(1):10. DOI: 10.3390/pr7010010

[21] Bravo J, Zhai L, Wu Z, Cohen RE, Rubner MF. Transparent superhydrophobic films based on silica nanoparticles. Langmuir. 2007;**23**(13):7293-7298. DOI: 10.1021/la070159q

[22] Li S, Xie H, Zhang S, Wang X. Facile transformation of hydrophilic cellulose into superhydrophobic cellulose. Chemical Communications. 2007;**1**(46):4857-4859. DOI: 10.1039/b712056g

[23] Song J, Rojas OJ. PAPER CHEMISTRY: Approaching super-hydrophobicity from cellulosic materials: A review. Nordic Pulp & Paper Research Journal. 2018;**28**(2):216-238

[24] Teisala H, Tuominen M, Kuusipalo J. Superhydrophobic coatings on cellulose-based materials: Fabrication, properties, and applications. Advanced Materials Interfaces. 2014;**1**(1):1-20. DOI: 10.1002/admi.201300026

[25] Manca M, Cannavale A, De Marco L, Aricò AS, Cingolani R, Gigli G. Durable superhydrophobic and antireflective surfaces by trimethylsilanized silica nanoparticles-based sol-gel processing. Langmuir. 2009;**25**(11):6357-6362. DOI: 10.1021/la804166t

[26] Salam A, Lucia LA, Jameel H. Fluorine-based surface decorated cellulose nanocrystals as potential hydrophobic and oleophobic materials. Cellulose. 2015;**22**(1):397-406. DOI: 10.1007/s10570-014-0507-9

[27] Shang W, Huang J, Luo H, Chang PR, Feng J, Xie G. Hydrophobic modification of cellulose nanocrystal via covalently grafting of castor oil. Cellulose. 2013;**20**(1):179-190. DOI: 10.1007/s10570-012-9795-0

[28] Salajková M, Berglund LA, Zhou Q. Hydrophobic cellulose nanocrystals modified with quaternary ammonium salts. Journal of Materials Chemistry. 2012;**22**(37):19798-19805. DOI: 10.1039/c2jm34355j

[29] Yu M, Gu G, Meng WD, Qing FL. Superhydrophobic cotton fabric coating based on a complex layer of silica nanoparticles and perfluorooctylated quaternary ammonium silane coupling agent. Applied Surface Science. 2007;**253**(7):3669-3673. DOI: 10.1016/j.apsusc.2006.07.086

[30] Ogihara H, Xie J, Okagaki J, Saji T. Simple method for preparing Superhydrophobic paper: Spray-deposited hydrophobic silica nanoparticle coatings exhibit high water-repellency and transparency. Langmuir. 2012;**28**(10):4605-4608

[31] Mauriello G, De Luca E, La Storia A, Villani F, Ercolini D. Antimicrobial

activity of a nisin-activated plastic film for food packaging. Letters in Applied Microbiology. 2005;**41**(6):464-469. DOI: 10.1111/j.1472-765X.2005.01796.x

[32] Kodjak A. FDA Bans 19 Chemicals Used in Antibacterial Soaps. New York, USA: Food and Drug Administration; 2016

[33] Laxminarayan R et al. Antibiotic resistance-the need for global solutions. The Lancet Infectious Diseases. 2013;**13**(12):1057-1098. DOI: 10.1016/S1473-3099(13)70318-9

[34] Wang L, Hu C, Shao L. The antimicrobial activity of nanoparticles: Present situation and prospects for the future. International Journal of Nanomedicine. 2017;**12**:1227-1249

[35] Lü F, Ye X, Liu D. Review of antimicrobial food packaging. Nongye Jixie Xuebao/Transactions of the Chinese Society of Agricultural Machinery. 2009;**40**(6):138-142

[36] Huang KS, Yang CH, Huang SL, Chen CY, Lu YY, Lin YS. Recent advances in antimicrobial polymers: A mini-review. International Journal of Molecular Sciences. 2016;**17**(9):578. DOI: 10.3390/ijms17091578

[37] Fortunati E et al. Nano-biocomposite films with modified cellulose nanocrystals and synthesized silver nanoparticles. Carbohydrate Polymers. 2014;**101**(1):1122-1133. DOI: 10.1016/j.carbpol.2013.10.055

[38] Kreuter J. Nanoparticles and microparticles for drug and vaccine delivery. Journal of Anatomy. 1996;**189**(Pt 3):503-505. DOI: 10.1002/bit

[39] Lam E, Male KB, Chong JH, Leung ACW, Luong JHT. Applications of functionalized and nanoparticle-modified nanocrystalline cellulose. Trends in Biotechnology. 2012;**30**(5):283-290. DOI: 10.1016/j.tibtech.2012.02.001

[40] Ramos AI et al. Analysis of the microcrystalline inclusion compounds of triclosan with β-cyclodextrin and its tris-O-methylated derivative. Journal of Pharmaceutical and Biomedical Analysis. 2013;**80**:34-43. DOI: 10.1016/j.jpba.2013.02.033

[41] Suller MTE, Russell AD. Triclosan and antibiotic resistance in *Staphylococcus aureus*. Journal of Antimicrobial Chemotherapy. 2000;**46**(1):11-18. DOI: 10.1093/jac/46.1.11

[42] Helander IM, Nurmiaho-Lassila EL, Ahvenainen R, Rhoades J, Roller S. Chitosan disrupts the barrier properties of the outer membrane of gram-negative bacteria. International Journal of Food Microbiology. 2001;**71**(2-3):235-244. DOI: 10.1016/S0168-1605(01)00609-2

[43] Rabea EI, Badawy MET, Stevens CV, Smagghe G, Steurbaut W. Chitosan as antimicrobial agent: Applications and mode of action. Biomacromolecules. 2003;**4**(6):1457-1465. DOI: 10.1021/bm034130m

[44] Cha DS, Chinnan MS. Biopolymer-based antimicrobial packaging: A review. Critical Reviews in Food Science and Nutrition. 2004;**44**(4):223-237. DOI: 10.1080/10408690490464276

[45] Panchal P et al. Graft modification of cellulose nanocrystals: Via nitroxide-mediated polymerisation. Biomacromolecules. 2014;**1**(1):235-244. DOI: 10.1021/la070159q

[46] Lee CH, Park HJ, Lee DS. Influence of antimicrobial packaging on kinetics of spoilage microbial growth in milk and orange juice. Journal of Food

Engineering. 2004;**65**(4):527-531. DOI: 10.1016/j.jfoodeng.2004.02.016

[47] El-Tahlawy KF, El-Bendary MA, Elhendawy AG, Hudson SM. The antimicrobial activity of cotton fabrics treated with different crosslinking agents and chitosan. Carbohydrate Polymers. 2005;**60**(4):421-430. DOI: 10.1016/j.carbpol.2005.02.019

[48] Weishaupt R et al. Enhanced antimicrobial activity and structural transitions of a nanofibrillated cellulose-nisin biocomposite suspension. ACS Applied Materials and Interfaces. 2018;**10**(23):20170-20181. DOI: 10.1021/acsami.8b04470

[49] Salmieri S et al. Antimicrobial nanocomposite films made of poly(lactic acid)–cellulose nanocrystals (PLA–CNC) in food applications—Part B: Effect of oregano essential oil release on the inactivation of listeria monocytogenes in mixed vegetables. Cellulose. 2014;**21**(6):4271-4285. DOI: 10.1007/s10570-014-0406-0

[50] Zhang H, Chen H, She Y, Zheng X, Pu J. Anti-yellowing property of polyurethane improved by the use of surface-modified nanocrystalline cellulose. BioResources. 2014;**9**(1):673-684. DOI: 10.15376/biores.9.1.673-684

[51] Ma IAW, Sh A, Ramesh K, Vengadaesvaran B, Ramesh S, Arof AK. Anticorrosion properties of epoxy-nanochitosan nanocomposite coating. Progress in Organic Coatings. 2017;**113**(2008):74-81. DOI: 10.1016/j.porgcoat.2017.08.014

[52] Panda SS, Katz HE, Tovar JD. Solid-state electrical applications of protein and peptide based nanomaterials. Chemical Society Reviews. 2018;**47**(10):3640-3658. DOI: 10.1039/c7cs00817a

[53] Csoka L, Hoeger IC, Rojas OJ, Peszlen I, Pawlak JJ, Peralta PN. Piezoelectric effect of cellulose nanocrystals thin films. ACS Macro Letters. 2012;**1**(7):867-870. DOI: 10.1021/mz300234a

[54] Csoka L, Hoeger IC, Peralta P, Peszlen I, Rojas OJ. Dielectrophoresis of cellulose nanocrystals and alignment in ultrathin films by electric field-assisted shear assembly. Journal of Colloid and Interface Science. 2011;**363**(1):206-212. DOI: 10.1016/j.jcis.2011.07.045

[55] Le Bras D, Strømme M, Mihranyan A. Characterization of dielectric properties of nanocellulose from wood and algae for electrical insulator applications. Journal of Physical Chemistry B. 2015;**119**(18):5911-5917. DOI: 10.1021/acs.jpcb.5b00715

[56] Mihranyan A, Llagostera AP, Karmhag R, Strømme M, Ek R. Moisture sorption by cellulose powders of varying crystallinity. International Journal of Pharmaceutics. 2004;**269**(2):433-442. DOI: 10.1016/j.ijpharm.2003.09.030

[57] Gaspar D et al. Nanocrystalline cellulose applied simultaneously as the gate dielectric and the substrate in flexible field effect transistors. Nanotechnology. 2014;**25**(9):094008. DOI: 10.1088/0957-4484/25/9/094008

[58] Pereira L, Gaspar D, Guerin D, Delattre A, Fortunato E, Martins R. The influence of fibril composition and dimension on the performance of paper gated oxide transistors. Nanotechnology. 2014;**25**(9):094007. DOI: 10.1088/0957-4484/25/9/094007

[59] Liu W, Ullah B, Kuo C-C, Cai X. Two-dimensional nanomaterials-based polymer composites: Fabrication and energy storage applications. Advances in Polymer Technology. 2019;**2019**:1-15. DOI: 10.1155/2019/4294306

[60] Tang L, Zhao L, Guan L. 7 graphene/polymer composite materials: Processing, properties and applications. In: Advanced Composite Materials: Properties and Applications. Berlin, Germany: De Gruyter Open; 2017

[61] Kiziltas A, Erbas Kiziltas E, Boran S, Gardner DJ. Micro-and nanocellulose composites for automotive applications. In: Society of Plastics Engineers -13th Annual Automotive Composites Conference and Exhibition, ACCE 2013, no. February 2016. Vol. 1. Novi, MI, USA: Proceedings of the SPE Automotive Composites Conference and Exhibition (ACCE); 2013. pp. 402-414

[62] Sanoj P, Balasubramanian K. High performance structural nano cellulose composites for motor vehicle spring suspension system. International Journal of Plastics Technology. 2014;**18**:383-389. DOI: 10.1007/s12588-014-9098-4

[63] Reshmy R et al. Nanocellulose-based products for sustainable applications-recent trends and possibilities. Reviews in Environmental Science and Biotechnology. 2020;**19**(4):779-806. DOI: 10.1007/s11157-020-09551-z

[64] Bordes P, Pollet E, Avérous L. Nano-biocomposites: Biodegradable polyester/nanoclay systems. Progress in Polymer Science (Oxford). 2009;**34**(2):125-155. DOI: 10.1016/j.progpolymsci.2008.10.002

[65] Xu S, Girouard N, Schueneman G, Shofner ML, Meredith JC. Mechanical and thermal properties of waterborne epoxy composites containing cellulose nanocrystals. Polymer. 2013;**54**(24):6589-6598. DOI: 10.1016/j.polymer.2013.10.011

[66] Bras J, Hassan ML, Bruzesse C, Hassan EA, El-Wakil NA, Dufresne A. Mechanical, barrier, and biodegradability properties of bagasse cellulose whiskers reinforced natural rubber nanocomposites. Industrial Crops and Products. 2010;**32**(3):627-633. DOI: 10.1016/j.indcrop.2010.07.018

[67] Cao Y, Zavaterri P, Youngblood J, Moon R, Weiss J. The influence of cellulose nanocrystal additions on the performance of cement paste. Cement and Concrete Composites. 2015;**56**:73-83. DOI: 10.1016/j.cemconcomp.2014.11.008

[68] Jorfi M, Foster EJ. Recent advances in nanocellulose for biomedical applications. Journal of Applied Polymer Science. 2015;**132**(14):1-19. DOI: 10.1002/app.41719

[69] Dong S, Roman M. Fluorescently labeled cellulose nanocrystals for bioimaging applications. Journal of the American Chemical Society. 2007;**129**(45):13810-13811. DOI: 10.1021/ja076196l

[70] Kilgore MB et al. Cell-based analysis of the immune and antioxidant response of the nanocarrier β -cyclodextrin conjugated with cellulose nanocrystals mechanistic insights on the immune and antioxidant response of functionalized cellulose nanocrystals: Does surface charge. Free Radical Biology and Medicine. 2018;**128**:S102

[71] Domingues RMA, Gomes ME, Reis RL. The potential of cellulose nanocrystals in tissue engineering strategies. Biomacromolecules. 2014;**15**(7):2327-2346. DOI: 10.1021/bm500524s

[72] Yang X, Cranston ED. Chemically cross-linked cellulose nanocrystal aerogels with shape recovery and superabsorbent properties. Chemistry of Materials. 2014;**26**(20):6016-6025. DOI: 10.1021/cm502873c

[73] Zhou Y et al. Recyclable organic solar cells on cellulose nanocrystal substrates. Scientific Reports. 2013;**3**:24-26. DOI: 10.1038/srep01536

[74] Kim JH et al. Review of nanocellulose for sustainable future materials. International Journal of Precision Engineering and Manufacturing-Green Technology. 2015;**2**(2):197-213. DOI: 10.1007/s40684-015-0024-9

[75] Qiu X, Hu S. 'Smart' materials based on cellulose: A review of the preparations, properties, and applications. Materials. 2013;**6**(3):738-781. DOI: 10.3390/ma6030738

[76] Way AE, Hsu L, Shanmuganathan K, Weder C, Rowan SJ. PH-responsive cellulose nanocrystal gels and nanocomposites. ACS Macro Letters. 2012;**1**(8):1001-1006. DOI: 10.1021/mz3003006

[77] Kafy A, Akther A, Shishir MIR, Kim HC, Yun Y, Kim J. Cellulose nanocrystal/graphene oxide composite film as humidity sensor. Sensors and Actuators, A: Physical. 2016;**247**:221-226. DOI: 10.1016/j.sna.2016.05.045

[78] Zhang L, Li Q, Zhou J, Zhang L. Synthesis and photophysical behavior of pyrene-bearing cellulose nanocrystals for Fe 3+ sensing. Macromolecular Chemistry and Physics. 2012;**213**(15):1612-1617. DOI: 10.1002/macp.201200233

[79] Sadasivuni KK, Kafy A, Zhai L, Ko HU, Mun S, Kim J. Transparent and flexible cellulose nanocrystal/reduced graphene oxide film for proximity sensing. Small. 2015;**11**(8):994-1002. DOI: 10.1002/smll.201402109

[80] Wang S, Zhang X, Wu X, Lu C. Tailoring percolating conductive networks of natural rubber composites for flexible strain sensors via a cellulose nanocrystal templated assembly. Soft Matter. 2016;**12**(3):845-852. DOI: 10.1039/c5sm01958c

[81] C. A. de Assis et al., "Conversion economics of Forest biomaterials: Risk and financial analysis of CNC manufacturing," Biofuels, Bioproducts and Biorefining, vol. 11, no. 4, pp. 682-700, Jul. 2017, doi: doi:10.1002/bbb.1782.

[82] Jiang G et al. Bacterial nanocellulose/Nafion composite membranes for low temperature polymer electrolyte fuel cells. Journal of Power Sources. 2015;**273**:697-706. DOI: 10.1016/j.jpowsour.2014.09.145

[83] M. J. Blair and W. E. Mabee, "Techno-economic and market analysis of two emerging forest biorefining technologies," Biofuels, Bioproducts and Biorefining, vol. 15, no. 5, pp. 1301-1317, Sep. 2021, doi: doi:10.1002/bbb.2218.

[84] Bondancia TJ et al. Production of nanocellulose using citric acid in a biorefinery concept: Effect of the hydrolysis reaction time and techno-economic analysis. Industrial and Engineering Chemistry Research. 2020;**59**(25):11505-11516. DOI: 10.1021/acs.iecr.0c01359

[85] Shen R et al. Research Progress and development demand of nanocellulose reinforced polymer composites. Polymers (Basel). 2020;**12**(9):2113. DOI: 10.3390/polym12092113

[86] Arantes V, Dias IKR, Berto GL, Pereira B, Marotti BS, Nogueira CFO. The current status of the enzyme-mediated isolation and functionalization of nanocelluloses: Production, properties, techno-economics, and opportunities. Cellulose. 2020;**27**(18):10571-10630. DOI: 10.1007/s10570-020-03332-1

[87] Squinca P, Bilatto S, Badino AC, Farinas CS. Nanocellulose production in future biorefineries: An integrated approach using tailor-made enzymes. ACS Sustainable Chemistry & Engineering. 2020;**8**(5):2277-2286. DOI: 10.1021/acssuschemeng.9b06790

[88] Jiang J, Zhu Y, Jiang F. Sustainable isolation of nanocellulose from cellulose and lignocellulosic feedstocks: Recent progress and perspectives. Carbohydrate Polymers. 2021;**267**:118188. DOI: 10.1016/j.carbpol.2021.118188

[89] Kaur P et al. Nanocellulose: Resources, physio-chemical properties, current uses and future applications. Frontiers in Nanotechnology. 2021;**3**:747329. DOI: 10.3389/fnano.2021.747329

[90] Stoudmann N, Nowack B, Som C. Prospective environmental risk assessment of nanocellulose for Europe. Environmental Science: Nano. 2019;**6**(8):2520-2531. DOI: 10.1039/C9EN00472F

[91] Piccinno F, Hischier R, Seeger S, Som C. Predicting the environmental impact of a future nanocellulose production at industrial scale: Application of the life cycle assessment scale-up framework. Journal of Cleaner Production. 2018;**174**:283-295. DOI: 10.1016/j.jclepro.2017.10.226

[92] Turk J, Oven P, Poljanšek I, Lešek A, Knez F, Malovrh Rebec K. Evaluation of an environmental profile comparison for nanocellulose production and supply chain by applying different life cycle assessment methods. Journal of Cleaner Production. 2020;**247**:119107. DOI: 10.1016/j.jclepro.2019.119107

Chapter 2

Spray-Coated Cellulose Nanofiber Films: Preparation, Characterization and Application

Kirubanandan Shanmugam

Abstract

Spraying cellulose nanofibers on the polished stainless-steel plate is a novel approach for the fabrication of free-standing cellulose nanofiber film (CNF). Recently, free-standing cellulose nanofiber film has gained attention as an alternative to synthetic plastic film. Free-standing/self-standing CNF film can be used as a potential barrier, in packaging application, as membranes for waste water application, in fabrication of biomedical film for wound repair and many more such applications in the fabrication of functional materials. To hasten the production of free-standing CNF film, spraying process is a considerable process-intensified method for large-scale production of film in a rapid manner. Spraying CNF on the stainless-steel plate produces the film with unique surfaces, namely a rough surface exposed to air and a smooth surface from the steel surface. The smooth surface of the film is very shiny and glossy and provides a platform for utilizing this smoothness for fabricating the functional materials such as a base substrate for flexible electronics and solar cells, etc. This chapter summarizes the production of free-standing CNF film via spraying and its characterization linked to its application.

Keywords: spraying, cellulose nanofibers, free-standing films, air permanence, uniformity, thickness mapping, water vapor permeability

1. Introduction

Plastic pollution is one of the serious threats to the environment in the current scenario. Packaging is the main source for these plastics' pollution and it is an alarming status to replace the synthetic plastics with biopolymers. Biopolymers are good alternatives for synthetic plastics, as they have biodegradability, serve as eco-friendly material, and possess good mechanical and barrier properties for the development of various functional materials [1]. Recently, cellulose nanofiber (CNF) has been getting a predominant place in the list of biopolymers [2].

Cellulose is the most important biorenewable, biodegradable and biopolymer that is available in plenty in nature and plays an excellent feed stock for the development of various sustainable materials on an industrial-scale production [3]. From the past decade, cellulose nanofiber is used as one of the pioneering feed stocks for the development of various functional materials. It is produced by the disintegration and delamination of cellulose fibrils from pulp that is, in turn, produced from a variety of green sources like wood, potato tuber, hemp and flax. It has a dimension diameter ranging from 5 to 60 nanometers (nm) and length of several micrometers [4]. Moreover, having a smaller dimension in cellulose nanofibrils in CNF gives the larger surface of CNF, which is why, there is a great opportunity for developing more functional materials for various applications [4, 5].

The films made from cellulose nanofiber (CNF) have various outstanding mechanical, optical and structural properties and these properties are played to fabricate various functional materials, such as cellulose nanocomposite [4], microfibrillated film [6], inorganic nanocomposite [7], organic transistors and conducting materials [8] and immunoassays and diagnostic materials [9]. Moreover, another advantage of the nanofibrillated cellulose is that it is easy to tailor its surface properties and mechanical properties. As a result, it is used in the field of photonics, surface modifications, nanocomposites, biomedical scaffold and optoelectronics [10]. Due to barrier and colloidal properties of CNF, it is widely used in paper-making, packaging and coatings to enhance its barrier surface and then in automotive industries [11]. Recently, CNF sheets turned out to be one of the most promising high-performance functional materials potentially used as filters [12], adsorbents, catalysts [13], cell culture substrates, thermal insulators and drug carriers [14].

On top of abovementioned excellent properties, these materials are biodegradable and recyclable. Hence, they have the potential to replace some of the synthetic polymeric materials that cause serious environmental problems [1]. However, persisting problems faced during CNF film preparation include low-energy consumption and rapidity in fabrication [15]. CNF films were prepared by vacuum filtration, casting and spray coating. However, these methods are time-consuming processes, and the films are low in weight and thickness [15, 16]. Even though these processes could be efficient in the way of producing better quality of films, they have constraints in the scaling-up process in large-scale production for technology transfer and commercialization of the free-standing films [15].

In the vacuum filtration method, the CNF film formation required a high dewatering time, which shows up as a major constraint for an industrial-scale process. Furthermore, Varanasi and Batchelor [17] reported on the rapid preparation of nanofibrillated sheet using a British hand sheet maker in 10 min. However, they achieved only the mass per unit area of 57.4 g/m^2 and thickness of 68.9 $\pm$ 8.90 μm. This is why the current investigation was motivated to develop a rapid and scalable spray coating technique to produce the nanocellulose film to replace the time-consuming conventional techniques for cellulose film [18]. Beneventi et al. [19] reported on the spray coating of the microfibrillated cellulose on the nylon fabric to prepare the nanopaper with maximum mass of the film of 124 g/m^2. In this approach, the experimental setup has a conveyor system with a speed of 0.5 m/min to achieve 124 g/m^2 of film's basis weight. However, it failed to explain the uniformity of the film through the thickness distribution and surface morphology. This chapter reveals the rapid preparation of a CNF film using a developed laboratory-scale spray coating system to produce high basis weight film in a short span of time.

2. Free-standing CNF film fabrication via spray coating process

Spraying cellulose nanofibers on the polished stainless-steel plate is a rapid and novel process for the fabrication of a CNF wet film [20]. Previously, spraying microfibrillated cellulose on three-dimensional (3D) brass metal was attempted to develop a film. However, the cracks and shrinkage on the film were formed. But this was the basic idea to develop a spray coating process to fabricate CNF film [21]. In the improved process, the microfibrillated cellulose suspension was sprayed on the fabric surface and then it was followed by vacuum filtration to remove the excess water present in the wet film [19]. The concept of spraying nanofibers has been generated from these concrete works. In addition to that, spraying provides contour coating and contactless coating with the solid surface. So that, the surface topography and morphology of the solid surface did not influence the coating process [15]. It has been reported that spraying process produces a CNF film with high basis weight without any change in the operation time [20]. The spraying CNF on the fabric surface and paper substrates was already developed for free-standing films and barrier coating on the papers' surface [19, 22]. This chapter reveals the spraying CNF on the stainless-steel plates for the fabrication of free-standing films.

The following experimental system on spray coating to fabricate CNF film was developed.

Figure 1 shows the experimental system for spraying CNF on the stainless-steel plate. It reveals the spraying CNF suspension on the metal plate that is kept on the conveyor. In this experimental system, there are two important parameters for tailoring the properties of a CNF film. CNF suspension consistency and velocity of the conveyor are the parameters used for tailoring the thickness and basis weight of the films. At constant velocity of the conveyor, the CNF suspension can be varied for spraying operation to get the CNF film. Normally, spraying low CNF% suspension produces the least thickness and basis weight of the film, whereas high basis weight and thickness of the film can be fabricated by spraying high CNF% suspension on the polished metal surface. Similarly, the CNF suspension concentration has been fixed and varying the velocity of the conveyor helps to tailor the thickness and basis weight of the film. It means that high basis weight and thickness of the film were fabricated at the lowest velocity of the conveyor. At that moment, a high amount of suspension gets

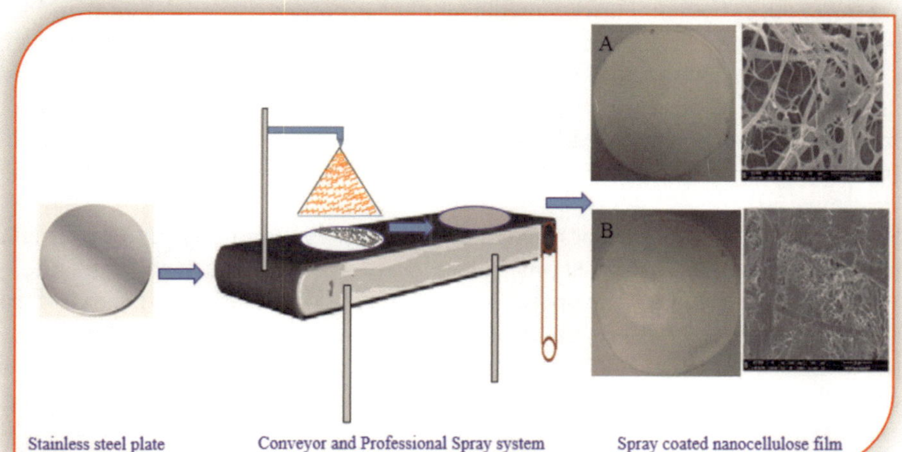

Figure 1.
Experimental setup for lab-scale spray coating system for the preparation of a nanocellulose (NC) film. (A) Rough surface of the NC film. (B) Smooth surface of the NC film.

deposited and fibers concentrated to form a high basis weight CNF film. Similarly, thin CNF films were fabricated spraying CNF suspension on the metal plate at high velocity of the conveyor. In this case, less amount of fibers were deposited on the steel plate to form a thin CNF film due to fast movement of the conveyor. Apart from these important parameters in a spray system to show their effect on CNF film's properties, the spray distance between the spray gun to the base surface, spray nozzle and spray gun position is indirectly controlling the film's properties, such as uniformity, thickness and basis weight [18, 20].

In this spray system, the CNF wet film can be fabricated and should be subjected to a drying process to remove the excess water in the wet film. The drying process can be carried out by different methods such as drying the wet film in an air oven at a temperature of 105°C and drying the wet sheets in a laminar flow chamber or fume hood with a constant flow of air under standard laboratory practice. The dried CNF film can be subjected to various characterizations and applications. The dried spray-coated CNF film has two unique but compact surfaces, namely rough surface and smooth surface. The rough surface is exposed to the air side when spraying CNF suspension on the metal plates. The smooth side of the CNF film is from the stainless-steel side and it has a shiny and glossy surface as one of the finishing qualities in this process. The surface smoothness of the CNF film is replicated from the stainless-steel plate. This smooth side of the film is used for the fabrication of numerous functional materials such as substrates for flexible electronics and printed electronics [16].

3. Analysis of spray-coated CNF films and their characterization

Spraying cellulose nanofiber suspension on the polished stainless-steel plates is a rapid process for the fabrication of a wet film of CNF. This method produces a compact film of cellulose nanofibrils having two unique surfaces, namely rough surface on the free side and smooth surface on the metal side. The operation time to form a 15.9-cm diameter CNF film consumes less than a minute and is independent of CNF suspension concentration. However, this method produces a wet CNF film and is subjected to a drying process to evaporate the water in a spray-coated CNF suspension. The drying of spray-coated wet film can be performed by keeping the wet film in an air oven at 105°C or air-drying in a laminar flow chamber under the standard laboratory conditions. The dried film on the stainless-steel plate can be easily peeled off from the plate and the CNF becomes free standing/self-standing films for various applications [20].

Figures 2 and **3** show the spray-coated CNF film. The spraying CNF on circular stainless-steel plate and square stainless-steel plate was achieved to fabricate the circular and square sheets. The operation time in spraying CNF suspension to form a 15.9-cm diameter film was less than a minute. Unlike vacuum filtration, the CNF concentration to fabricate film was independent of their operation time in a spraying process. The spraying CNF on the metal surface produces an ultrasmooth film for various applications [18].

Figure 4 reveals the cross-section of the scanning electron microscopy (SEM) micrographs of a CNF film prepared via spray coating. The SEM micrograph confirms the complex cellulose nanofibril layers intertwined through the hydrogen bonding between the hydroxyl groups of the CNF. This also increases tortuosity of the film and shows this effect on the barrier performance of the CNF film [16].

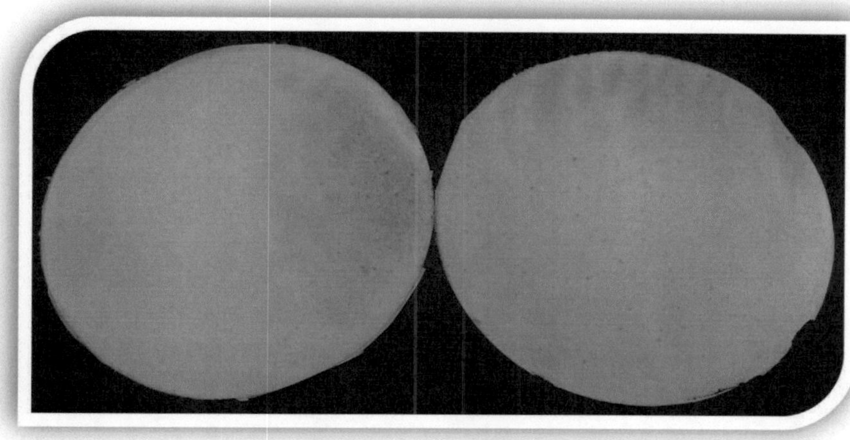

Figure 2.
Circular sheets of a CNF film via spray coating process.

Figure 3.
Square sheets of a CNF film via spray coating process.

Figures 5 and **6** show the SEM micrographs of the spray-coated CNF film and comparison with the CNF film prepared via vacuum filtration. The CNF film via spraying has compactness and roughness on the free side and smoothness on the other side. The rough surface of the film is very porous due to various sizes of the fibre distribution. The smooth side of the film has a shiny and glossy surface and its smoothness is replicated from the surface of the stainless-steel plate [18, 20]. The mechanism of replication of the smoothness from the stainless-steel plate remains obscure [16]. The rough and smooth surface of the spray-coated CNF film has an importance in the fabrication of various functional materials [16]. For example, in the construction of flexible electronics and printed electronics, the conductive ink on the cellulose substrates should be penetrated well on the surface of the substrates. To achieve this, the sufficient roughness/smoothness of the substrates are required for

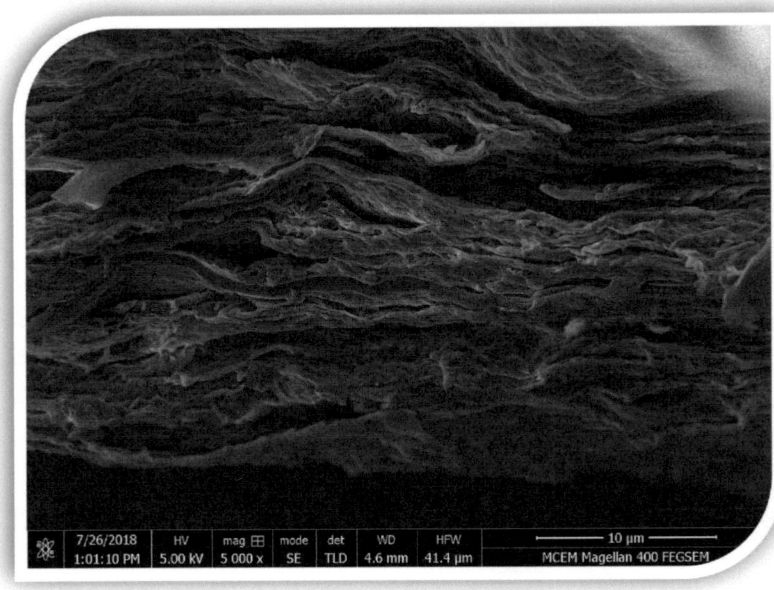

Figure 4.
Cross-sectional view of spray-coated CNF films.

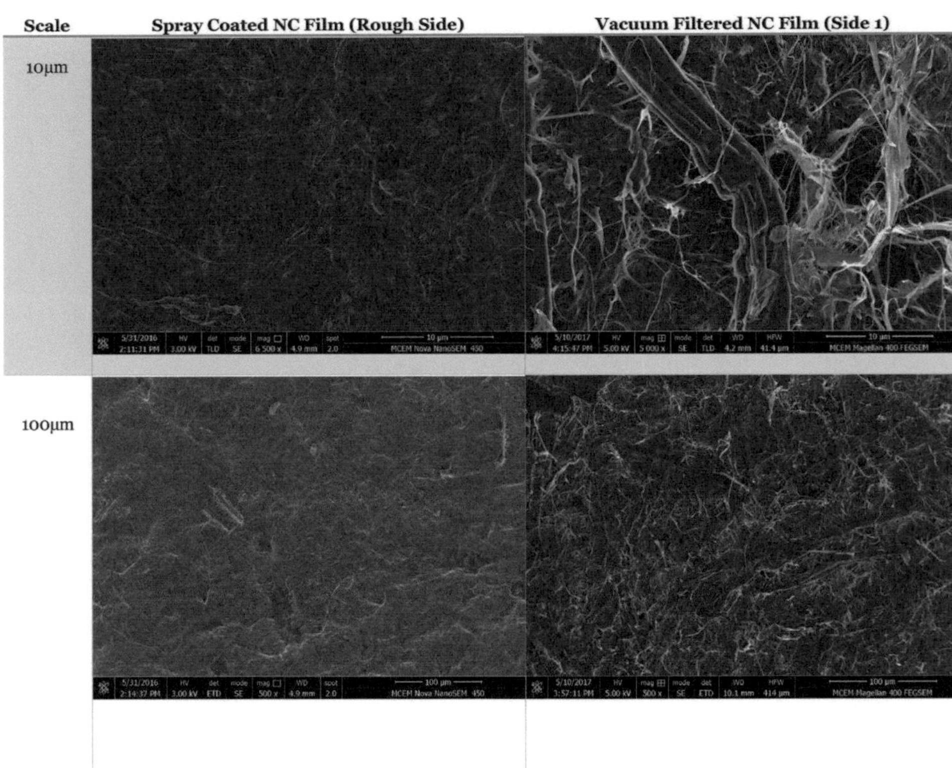

Figure 5.
Rough side of spray-coated CNF film and the filter side of the vacuum-filtered film.

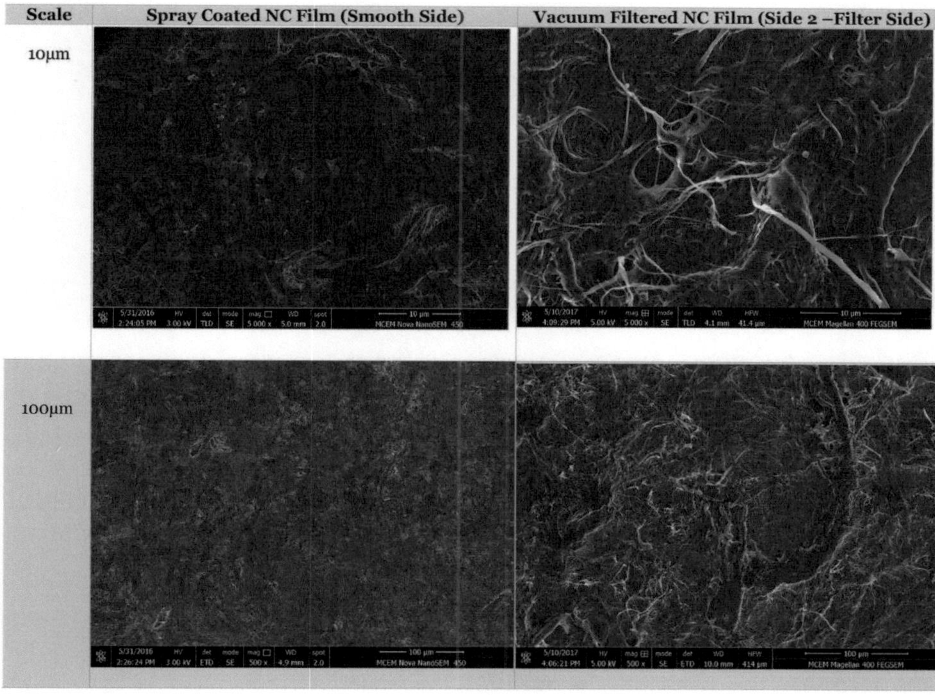

Figure 6.
Smooth side of a spray-coated CNF film and the free side of the vacuum-filtered film.

spreading the conductive ink. Similarly, the roughness and smoothness of the film can be used in the fabrication of solar cells [16].

The rough side of the spray-coated CNF film and the filter side of the vacuum-filtered CNF film are shown in **Figure 5**. The surfaces of both sides of the films are very porous and have a good surface roughness. **Figure 6** reveals the smooth side of the CNF film and the free side of the vacuum-filtered CNF film.

4. Surface roughness of a free-standing CNF film

As discussed earlier, the surface roughness of the film is one of the important criteria for the construction of functional materials. The surface roughness of the CNF film is evaluated by optical profilometry.

Figure 7 shows the optical profilometry image of the rough side of the CNF film prepared via spray coating. The rough side of the film was very porous and it showed high surface roughness on the film. This was due to various fiber size distributions in cellulose nanofibrils. The mean surface roughness on the rough side of the CNF film was found to be 1654 nm and the root mean square (RMS) value of the surface roughness on this side was reported to be 2087 nm. **Figure 8** reveals the optical profilometry image of the smooth side of a CNF film. The Ra and Rq values from the image confirm that the surface was very smooth and shiny and glossy. The Ra and Rq values from the image were evaluated to be 278 nm and 389 nm, respectively. **Figures 9** and **10** show the optical profilometry images of free and filter sides of the CNF film prepared via vacuum filtration. The Ra and Rq values on the free side of the filtered CNF film were evaluated

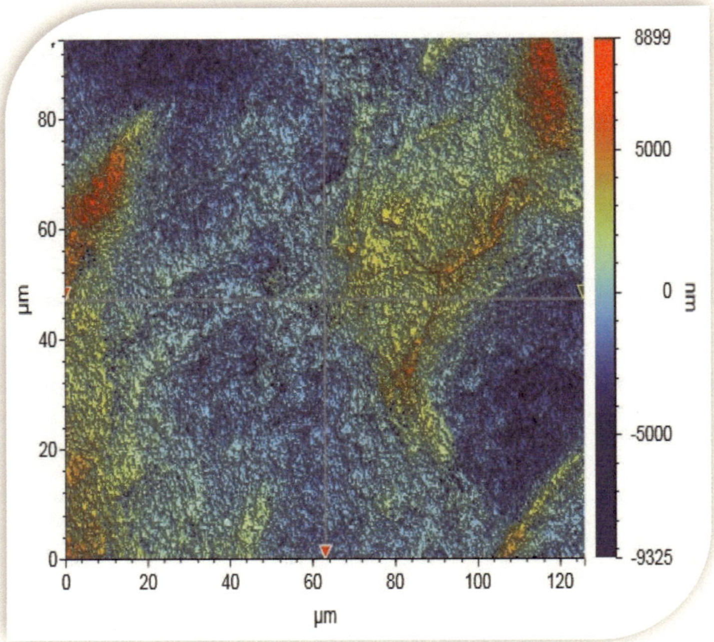

Figure 7.
Optical profilometry image of the rough side of a CNF film via spraying.

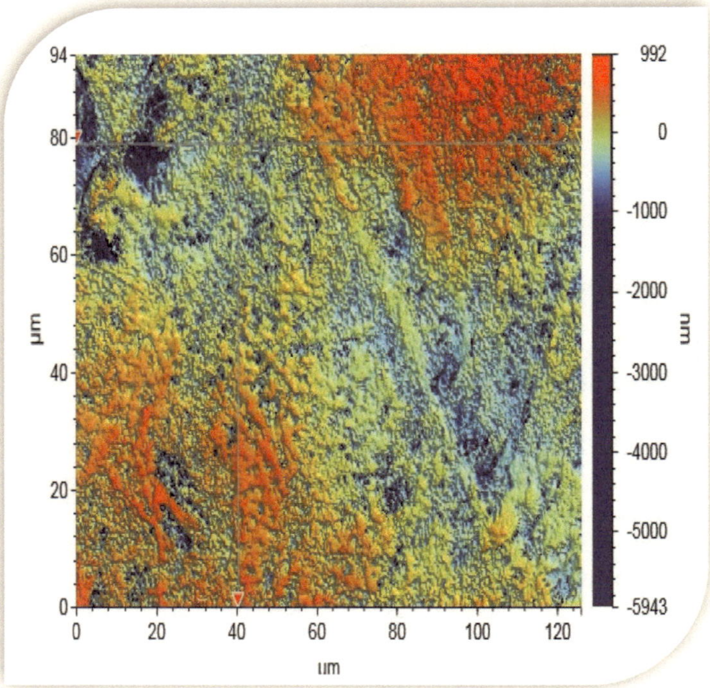

Figure 8.
Optical profilometry image of the smooth side of the CNF film via spraying.

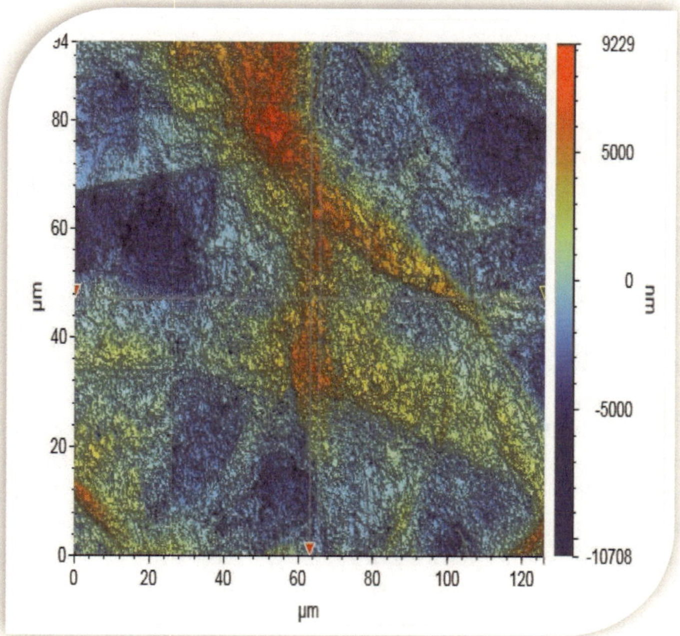

Figure 9.
Optical profilometry image of the free side of the CNF film via filtration.

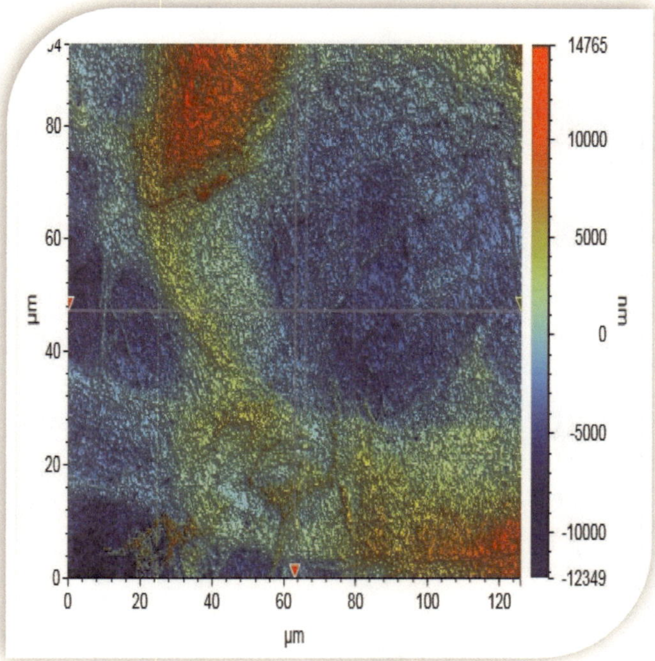

Figure 10.
Optical profilometry image of the filter side of the CNF film via filtration.

to be 2150 nm and 2673 nm, respectively. Similarly, the Ra and Rq values on the filter side of the vacuum-filtered CNF film were evaluated to be 3015 nm and 3751 nm, respectively. When comparing the surface roughness of the CNF film with that from filtration, the spray-coated CNF film has smooth and less porous surface [18, 20].

Spraying CNF suspension on the polished metal surface like stainless steel produces a more smooth film than that of standard procedure such as vacuum filtration. In the filtration process, the film has rough surface on the filter side and free side and is also porous, depending on the type of cellulose nanofiber used. In spraying, the film has unique surfaces, such as smooth side from the metal side and rough side from the air side.

5. Atomic force microscopy (AFM) studies

The AFM studies on the spray-coated CNF film were investigated to analyze the nanoscale surface, roughness of the rough and smooth surfaces of the film. The visual examination of the CNF film via spraying was rough on the free side of the film and smooth on the metal side. In addition to that, the smooth side was very shiny and glossy as one of the finishing qualities of the film. The RMS surface roughness of the CNF film from the AFM micrographs was evaluated to be 51.4 nm on the rough side and 16.7 nm on the smooth side in an inspection area of 2 μm × 2 μm. In the case of a vacuum-filtered CNF film, the RMS surface roughness of the CNF film was found to be 102.3 nm on the free side and 70.64 nm on the filter side on the same area of inspection. By this way, the nanoscale roughness of the CNF film was evaluated and it is implemented on these surfaces for the construction of printed and flexible electronics substrates. **Figures 11** and **12** show the surface roughness of the CNF film and RMS value evaluated from these AFM micrographs [16].

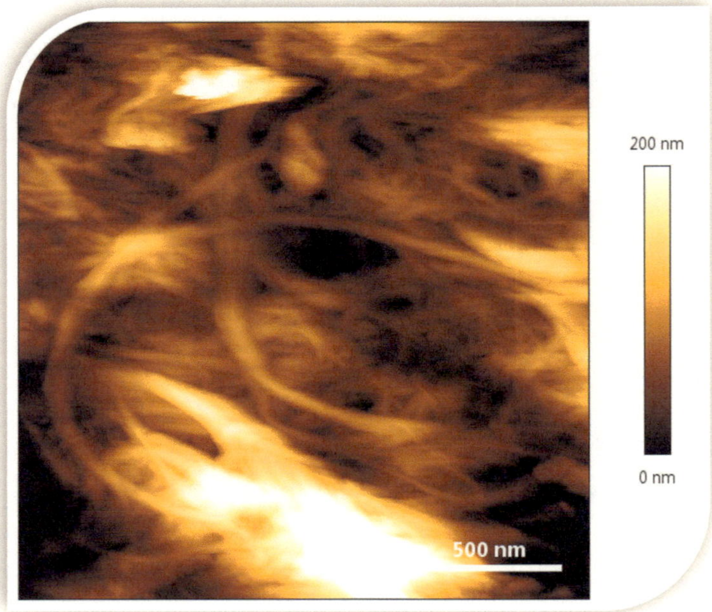

Figure 11.
Rough surface of the spray-coated CNF film.

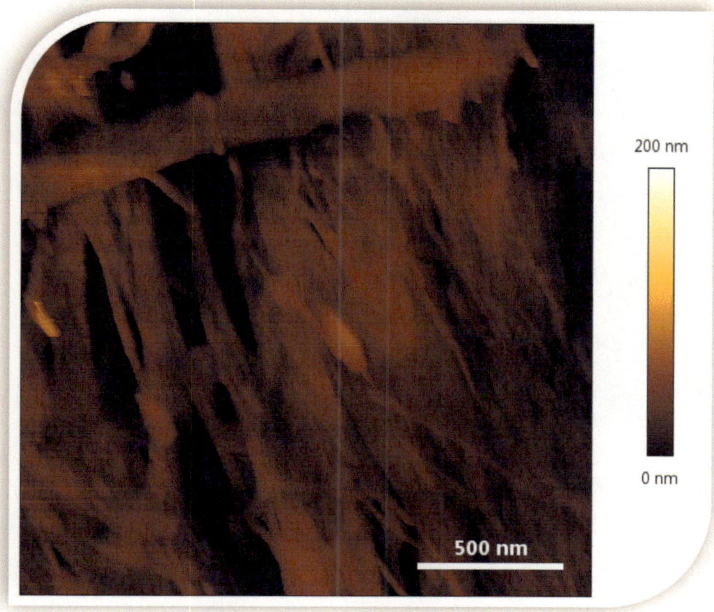

Figure 12.
Smooth surface of the CNF film via spraying.

6. Thickness mapping of CNF film via spraying

Thickness of the film is one of the main parameters for controlling the barrier performance of the film. **Figure 13** shows the thickness mapping of spray-coated CNF film and its comparison with that of a vacuum-filtered CNF film (**Figure 14**). The thickness mapping of the CNF film was evaluated from 1.5 wt.% CNF film via

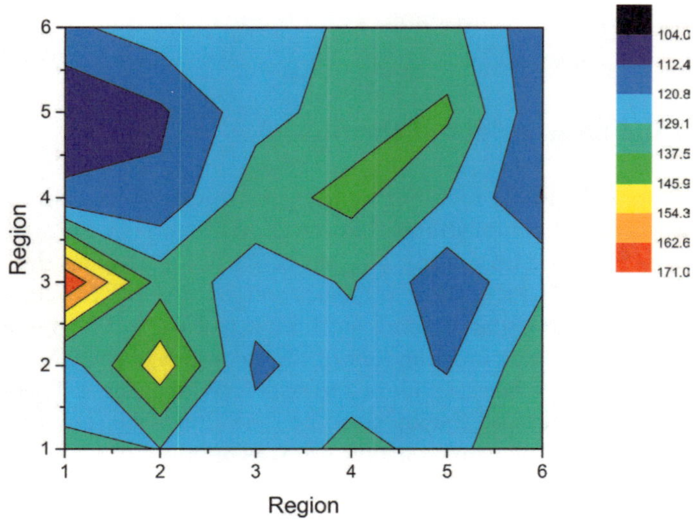

Figure 13.
Thickness mapping of the spray-coated CNF film.

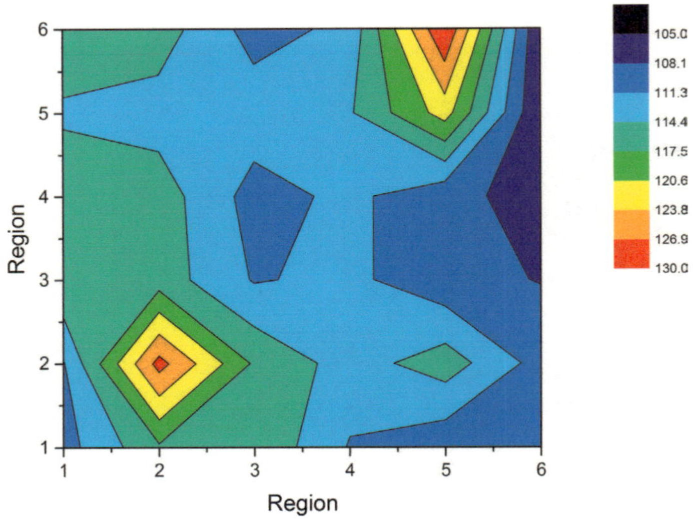

Figure 14.
Thickness mapping of the vacuum-filtered CNF film.

spraying to 1.5 wt.% CNF on the stainless-steel plate. The basis weights of the film produced by vacuum filtering and spray coating, respectively, are 100.5 ± 3.4 g/m^2 and 95.2 ± 5.2 g/m^2, respectively. Vacuum filtering consumes a substantially longer dewatering time of 15 minutes to form the film. In the spraying operation, the operation time to form a film is independent of CNF suspension concentration. Even after accounting for the little variation in basis weight, the spray-coated CNF film is somewhat thicker when compared to the vacuum-filtered film. The apparent densities of the vacuum-filtered and spray-coated films were 793 and 834 kg/m^3, respectively. Additionally, the thickness of the spray-coated film is distributed across a somewhat larger range. **Figures 13** and **14** reveal the uniform thickness of the CNF film fabricated via spraying and filtration processes. It seems that the spray-coated CNF film has better uniformity and is comparable with a filtered CNF film [20].

7. Thickness and basis weight of the film

Figure 15 reveals the linear relationship between thickness and basis weight of the CNF film via spraying. It demonstrates that the central composite design (CCD) model may be used to scale up the spraying process since it matches the actual experimental data well. The thickness and basis weight of the CNF film were tailored by increasing spraying CNF suspension from 1 wt.% to 2 wt.%. The operation time for spraying CNF suspension was independent of fiber content in the CNF suspension. The following models have been developed to scale up the process. These models reveal that the basis weight and thickness of the CNF film were found to be highly influenced by the CNF suspension concentration, as opposed to conveyor speed and spray distance based on the testing's findings.

$$\begin{aligned}\text{Basis weight} = -64.45 + 122.43 * \text{CNF suspension concentration} - 17.28 \\ * \text{Velocity of the conveyor} + 0.34 * \text{Spray distance}\end{aligned} \tag{1}$$

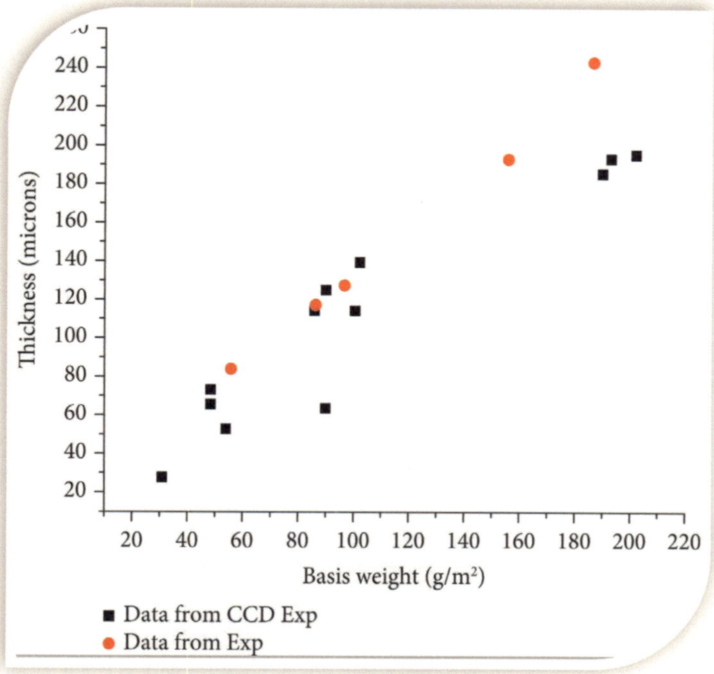

Figure 15.
Linear relationship between thickness and basis weight of the spray-coated CNF film.

$$\text{Thickness} = -0.106 + 0.111 * \text{CNF suspension concentration} - 0.000017 * \text{Velocity of the conveyor} + 0.002 * \text{Spray distance} \tag{2}$$

These are linear models confirming the direct relationship between thickness and basis weight of the CNF film.

From the linear models, there were two important parameters controlling the thickness and basis weight of the CNF film. Mainly, CNF suspension concentration and velocity of the conveyor in the experimental setup were deciding parameters for tailoring the CNF film's properties [23].

8. Mechanical performance of spray-coated CNF film

Figure 16 reveals the tensile index of the spray coated cellulose nanofiber film and its comparison with that of a vacuum filtration film. The modified configuration of an experimental spray system produces the CNF film that has tensile indices higher than that of the CNF film via vacuum filtration. This is because of the high uniformity of the CNF film that was fabricated in the modified configuration of the spray system. Generally, spraying CNF suspension on the polished metal plate is controlled by numerous parameters, mainly CNF suspension's consistency, process variables in the spray system, such as the spray distance, nozzle diameter, type of spray system and sprayability of the CNF suspension. These parameters indirectly control the uniformity of the film, which is linked with barrier and mechanical properties of the film [18].

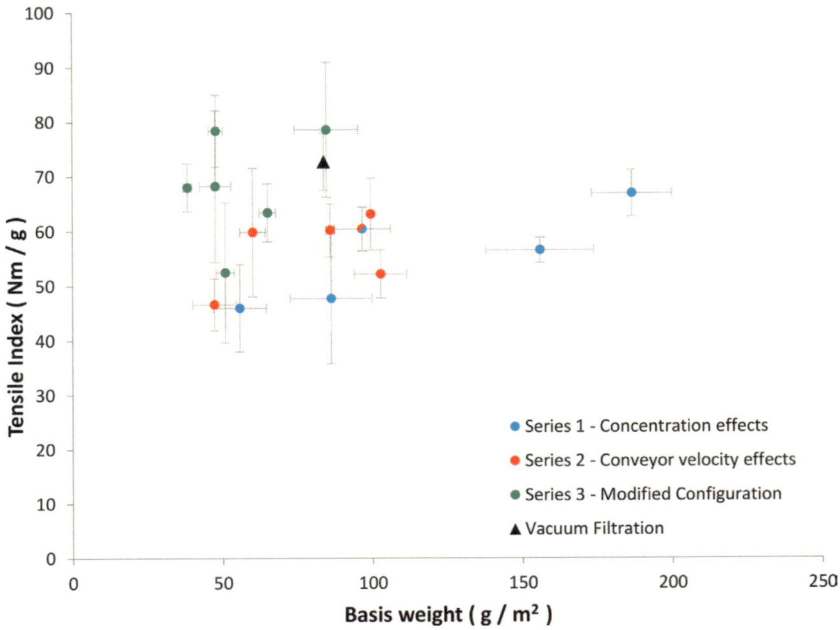

Figure 16.
Tensile index of the spray-coated cellulose nanofiber film and its comparison with that of a vacuum-filtered film.

9. Cost and environmental analysis

The free-standing CNF film has not been commercialized so far and it is in the researching and development (R&D) stage. The patents on free-standing CNF films and their composites produced via various methods have been increased. The cost of Diacel KY 100S from Diacel Chemical Pvt. Ltd. was 2500 Australian dollars (AUD) per 50 kg of nanocellulose. The cost of a spraying system was around 4000 AUD for the construction of an experimental system to spray CNF on the polished metal surface. The sizes of the film are 220 mm x 220 mm for square sheet and 159 mm for circular sheet. The basis weight of the film was assumed to be 100 g/m^2. For 1 kg of KY100S, 10 sheets can be fabricated and each sheet consumes 5 AUD for the fabrication. The operation and maintenance costs were not considered in this study. The operation time for the fabrication of a 220 mm × 220 mm square sheet and 159 mm circular sheet was less than a minute. When compared with vacuum filtration (VF), a laboratory version of the paper-making machine, spraying is a process intensified for the fabrication of free-standing cellulose nanofiber films and their composites. In the case of the spraying method, there are few steps, such as spraying CNF suspension on the metal plates, followed by drying under standard approach. It has required less fixed capital and operating cost and labour cost. In the case of vacuum filtration, there are many steps, such as the agitation of CNF suspension, mixing of CNF suspension, dewatering, couching, sheet removal, drying and then pressing. It confirms that filtration requires good fixed capital and high operating cost and labour cost. Spraying operation can be integrated with other coating methods such as roll to roll (R2R) for giving a high performance in the rate of production of a CNF film. So that, the cost of CNF film will be reduced for commercializing in the market.

Under controlled composting circumstances, the biodegradability and compostability of nanofibrillar cellulose-based (NFC) products, such as films, concentrated NFC, and paper products incorporating NFC, were assessed. All of the NFC products that were evaluated met the criteria for biodegradability outlined in European Standard EN 13432. NFC even increased the biodegradability of paper that had 1.5% NFC added to it. The modified pilot-scale composting test EN 14045 was used to assess disintegration during composting. In 3 weeks of composting, NFC films entirely decomposed, and NFC had no effect on how easily paper products containing NFC degraded. Using a bioluminescence test using *Vibrio fischeri*, ecotoxicity during the biodegradation of NFC products in a compost environment was assessed. For any of the samples, there was no evidence of acute toxicity [24].

10. Comparison with other coating process and validation

Spraying cellulose nanofibers on the polished stainless-steel plate is a rapid process to form a compact CNF film. It is a new process and should be compared with other conventional methods to confirm the efficacy of spray coating method [25]. The current spraying method has the capacity to handle the CNF suspension from 1.0 wt.% to 2.0 wt.% to produce the thickness and basis weight of CNF film from ~60 μm to ~200 μm and ~55 g/m^2 to ~199 g/m^2 [20]. The performance of the spray coating process can be improved by the high-performance spray system that can handle the CNF suspension of more than 2.00 wt.% [16, 18]. In addition to that, the rheology modifier, such as montmorillonite (MMT) clay, can be added into the CNF suspension for spraying to avoid any interruption in forming spray jet for fabrication of the films [16]. When comparing spraying microfribillated cellulose on the 3D structures [21], this method is quite reproducible and produces the CNF film without any cracks and homogeneous film with a high degree of uniformity [18]. The earlier method of spraying microfibrillar cellulose (MFC) on a 3D brass structure produces the MFC film having the basis weight and thickness from 59 to 118 g/m^2 and 46 to 68 μm, respectively. The spray-coated 3D structure consists of cracks and wrinkles formed on the surface. However, the reported method handled high solid MFC suspension, such as 4.5 wt.% and 9 wt.% MFC suspensions, resulting in the formation of a disturbed spray jet and also resulting in the formation of cracks and wrinkles on the film. A high MFC suspension behaves as a gel-like fluid and makes the spray systems lose the spraying ability [21].

Similar to the current spray coating method, the earlier spraying MFC on the nylon fabric was attempted to fabricate the free-standing CNF film. In this method, spraying MFC on the nylon fabric consumed time from 10 min to 20 min and then the wet sprayed film was subjected to water removal from the CNF suspension via applying vacuum, which is similar to vacuum filtration. The time consumed for filtration was from 15 sec to 90 sec and after that vacuum dried under standard temperature. The spray-coated MFC film from this spraying process has the basis weight that varied from 13.7 g/m^2 to 124 g/m^2, with the thickness of the film varying from 10 μm to 72 μm. It was also reported that the imprints of nylon fabric were marked on the spray-coated MFC film [19]. Spraying was more efficient in the fabrication of free-standing CNF film while compared with solvent casting, vacuum filtration and hot pressing. These processes were problematic in the evaporation or removal of solvent from CNF film and a time-consuming process and also limitation in the basis weight of the film. Spin coating is a laboratory approach for the fabrication of free-standing thin CNF film for the study of biomolecules' interaction. It is not a scalable method due to the removal of water from the suspension via spinning to form ultrathin films. This method can be used to coat the substrates for laboratory-scale studies. Roll-to-roll (R2R) coating is another approach for the fabrication of CNF films and capacity for large-scale production of the film. In this method, CNF was coated on the pre-treated substrates such as plastic films. The spreading of CNF on the substrates was a challenging task and then coated and dried under pressing after peeled from the substrates. The basis weight of the CNF film can be achieved from 0.1 to 400 g/m^2 [15].

Given this analysis, the spraying CNF on the base substrates is more advantageous in the fabrication of free-standing CNF film. Spraying on the stainless-steel plates produces the film with unique surfaces, mainly smooth on the steel side and it can be used for the fabrication of functional materials. When comparing with the other coating process, the operation time for spray coating in the current practice was less than a minute to fabricate a 15.9-cm diameter film and was independent of CNF suspension concentration. The integration of roll-to-roll coating with spray system is another approach for large-scale production of free-standing CNF films.

11. Application of spray-coated CNF films

Spray-coated CNF films have been utilized in various fields and applications as a substrate for developing functional materials. **Figure 17** shows various applications for CNF films in various fields. Due to the rapid process of spraying, it can be used as a barrier material to replace the synthetic plastics in the packaging sector. Generally, cellulose nanofibers are good as oxygen barrier and show a performance greater than that of synthetic plastics. However, the water vapor permeability (WVP) of CNF was not equalizing the water vapor barrier performance of synthetic plastics. Spraying CNF on the metal plates produces a film with compactness acting as a good barrier against water vapor and its performance was better than that of the vacuum filtered film and comparable with that of synthetic plastics. Furthermore, the water vapor

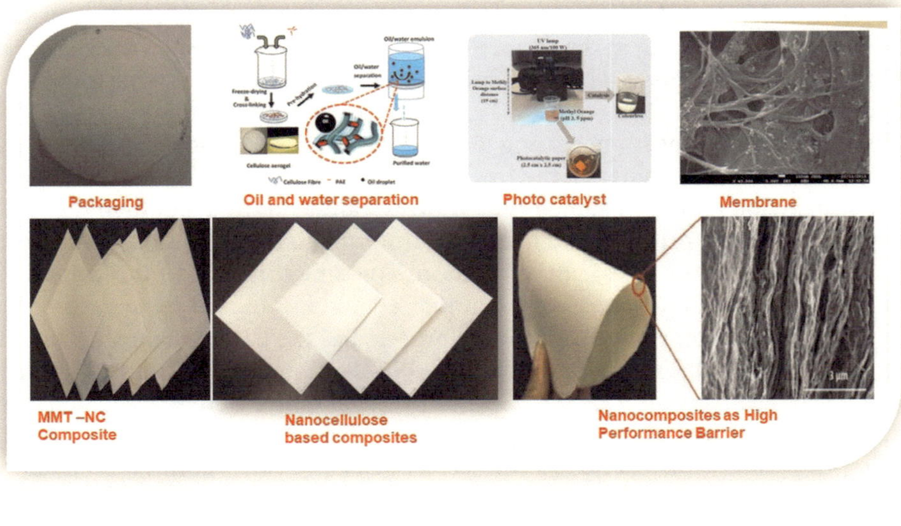

Figure 17.
Application of cellulose nanofiber films via spraying and filtration.

barrier of the CNF film was improved by incorporating nano-inorganics/antimicrobial inorganics into cellulose nanofibril matrix in the CNF suspension [15].

In the case of fabrication of nanocellulose-montmorillonite composite, the time taken for dewatering in vacuum filtration process was exponentially increased with MMT concentration and time was consumed from 3 hours to 24 hours, depending on the MMT content in cellulose nanofiber suspension. To mitigate this problem, spray coating has been implemented to fabricate CNF-MMT composite and here the operation time for spraying CNF-MMT suspension was independent of MMT concentration in CNF suspension. The spray-coated CNF MMT composite was a good barrier material for replacing synthetic plastics in the packaging materials. The antimicrobial inorganics incorporated into cellulose nanofiber suspension were fabricated as composite via spray coating process. This free-standing composite can be used in antimicrobial packaging and bioactive packaging. Similarly, the free-standing CNF film prepared via spraying can be used as the membrane for waste water treatment. In this composite, titanium dioxide was also impregnated into the film and its becomes a photocatalyst for waste water treatment application [15].

The membranes were also developed from spray-coated CNF film to separate oil and water mixture. In addition to that, various composites from CNF—inorganics can be fabricated via spray coating process for various applications. The spray-coated CNF films have unique surfaces, namely rough surface and smooth surface. The smooth surface of the CNF film can be used for the development of printed and flexible electronics. **Figure 18** shows the CNF film as substrates for printed electronics and flexible electronics [16].

The spray-coated CNF film can be used as a base biomaterial for the development of tissue engineering material and drug-delivery vehicle. The silver nanoparticle

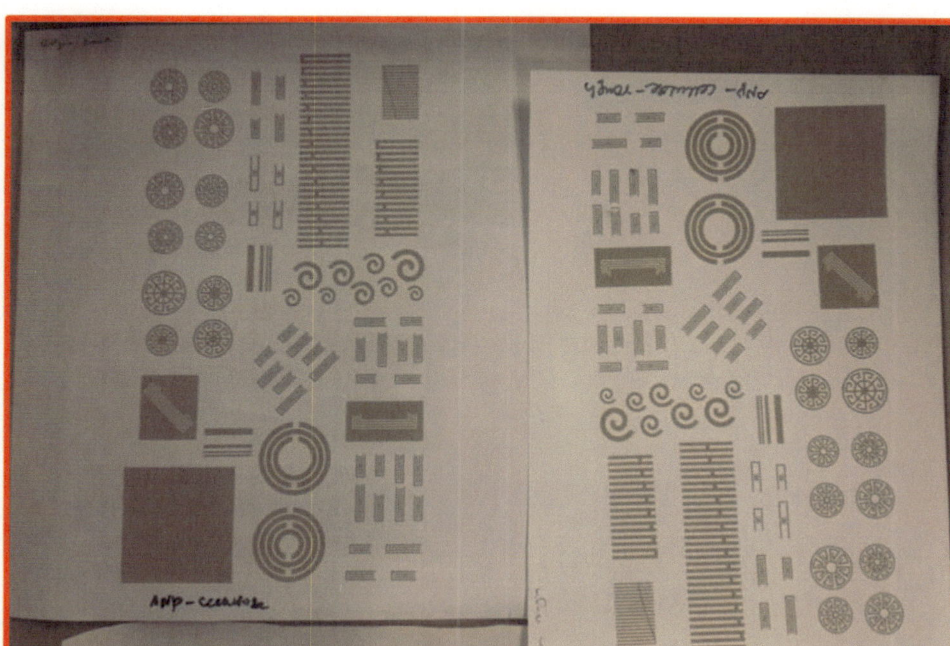

Figure 18.
The printed circuits on the spray-coated nanocellulose films.

(AgNP) and MMT were coated on the spray-coated CNF film via a laboratory spraying method to develop a drug-delivery vehicle composite for the treatment of wound. The silver nanoparticle present on the surface of CNF film can eradicate the wound pathogens at the wound site and the CNF film can act as a template for skin regeneration. In addition to spray coating to prepare free-standing CNF films and composites, this methodology can be used for developing CNF barrier layers on the paper and paper board substrates for enhancing their barrier potential against air and water vapor. Furthermore, the spraying CNF suspension was implemented to coat the CNF layers for membrane development for water treatment applications [15].

12. Spray-coated CNF film in packaging

The most important application of spraying CNF on the base surface was for the fabrication of CNF film, which can be used as a barrier material and as a good alternative for synthetic plastics. **Figure 19** demonstrates the spray-coated CNF film's capability as a reliable water vapor barrier and comparison to synthetic plastics. However, the barrier efficacy of the packing film against water vapor is also determined by its thickness. Because of this, the film's water vapor permeability—a number that was determined by normalizing the thickness of the film with its water vapor transmission rate (WVTR) values—was used to characterize the performance of the water vapor barrier. **Figure 12** compares synthetic plastics with spray-coated CNF film in terms of WVP. This graphic demonstrates how comparable the WVP of spray-coated CNF film is to that of synthetic polymers. Beyond this benefit, CNF is an environmentally benign nanomaterial with the ability to break down in the environment [25].

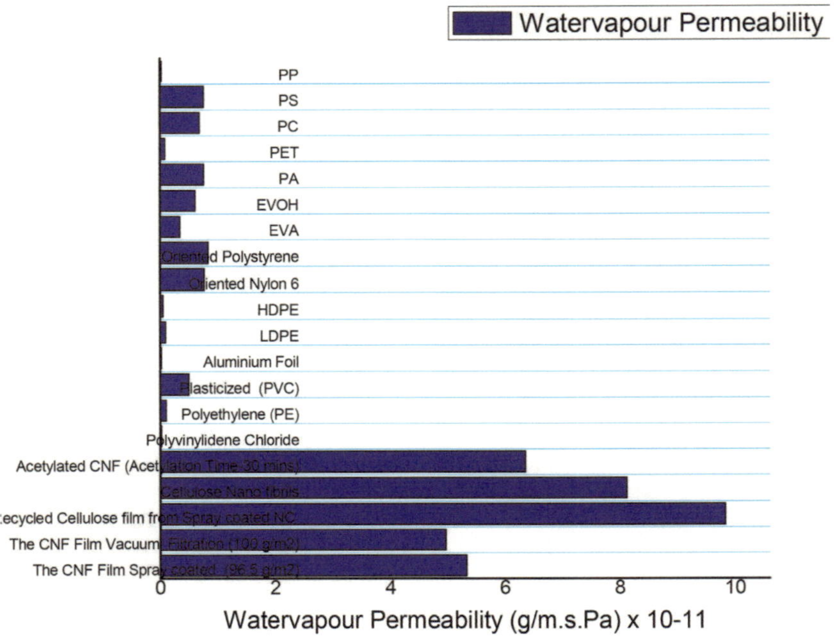

Figure 19.
Water vapor permeability of the CNF film and its comparison with that of synthetic plastics.

13. Conclusion

In order to meet the speed of production of film, a rapid process is required to fabricate the free-standing film of cellulose nanofibers/nanocellulose material. To answer this need, spraying/spray coating is a rapid process to fabricate the CNF film in a free-standing manner/self-standing sheets for various applications. The current spraying process produces the free-standing film in a rapid manner, taking up operation time less than a minute in forming the spray-coated wet film. However, the drying of a spray-coated wet film consumes a time of more than 24 hours in an air-drying process under standard laboratory conditions and a couple of hours in an oven drying at 105°C under standard practice. Unlike the vacuum filtration process, the operation time/film formation time of spraying process was independent of CNF suspension concentration and a potential for scaling up. The spray-coated film has unique surfaces, such as rough and smooth, and these surfaces lead to the development of various functional materials, such as packaging, membrane and drug-delivery devices.

Author details

Kirubanandan Shanmugam
Saveetha School of Engineering, SIMATS, Chennai, Tamilnadu, India

*Address all correspondence to: kirubanandan.shanmugam@gmail.com

IntechOpen

© 2024 The Author(s). Licensee IntechOpen. This chapter is distributed under the terms of the Creative Commons Attribution License (http://creativecommons.org/licenses/by/3.0), which permits unrestricted use, distribution, and reproduction in any medium, provided the original work is properly cited.

References

[1] Ncube LK, Ude AU, Ogunmuyiwa EN, Zulkifli R, Beas IN. Environmental impact of food packaging materials: A review of contemporary development from conventional plastics to polylactic acid based materials. Materials. 2020;**13**(21):4994

[2] Dufresne A. Nanocellulose: A new ageless bionanomaterial. Materials Today. 2013;**16**(6):220-227

[3] Klemm D, Kramer F, Moritz S, Lindström T, Ankerfors M, Gray D, et al. Nanocelluloses: A new family of nature-based materials. Angewandte Chemie International Edition. 2011;**50**(24): 5438-5466

[4] Henriksson M, Berglund LA, Isaksson P, Lindström T, Nishino T. Cellulose nanopaper structures of high toughness. Biomacromolecules. 2008; **9**(6):1579-1585

[5] Abe K, Iwamoto S, Yano H. Obtaining cellulose nanofibers with a uniform width of 15 nm from wood. Biomacromolecules. 2007;**8**:3276-3278

[6] Syverud K, Stenius P. Strength and barrier properties of MFC films. Cellulose. 2009;**16**:75-85

[7] Mörseburg K, Chinga-Carrasco G. Assessing the combined benefits of clay and nanofibrillated cellulose in layered TMP-based sheets. Cellulose. 2009;**16**: 795-806

[8] Chinga-Carrasco G, Tobjörk D, Österbacka R. Inkjet-printed silver nanoparticles on nano-engineered cellulose films for electrically conducting structures and organic transistors: Concept and challenges. Journal of Nanoparticle Research. 2012; **14**:1-10

[9] Orelma H, Filpponen I, Johansson LS, Österberg M, Rojas OJ, Laine J. Surface functionalized nanofibrillar cellulose (NFC) film as a platform for immunoassays and diagnostics. Biointerphases. 2012;**7**(1)

[10] Abitbol T, Rivkin A, Cao Y, Nevo Y, Abraham E, Ben-Shalom T, et al. Nanocellulose, a tiny fiber with huge applications. Current Opinion in Biotechnology. 2016;**39**:76-88

[11] Nechyporchuk O, Belgacem MN, Bras J. Production of cellulose nanofibrils: A review of recent advances. Industrial Crops and Products. 2016;**93**: 2-25

[12] Metreveli G, Wågberg L, Emmoth E, Belák S, Strømme M, Mihranyan A. A size-exclusion nanocellulose filter paper for virus removal. Advanced Healthcare Materials. 2014;**3**(10): 1546-1550

[13] Koga H, Tokunaga E, Hidaka M, Umemura Y, Saito T, Isogai A, et al. Topochemical synthesis and catalysis of metal nanoparticles exposed on crystalline cellulose nanofibers. Chemical Communications. 2010; **46**(45):8567-8569

[14] Bacakova L, Pajorova J, Bacakova M, Skogberg A, Kallio P, Kolarova K, et al. Versatile application of nanocellulose: From industry to skin tissue engineering and wound healing. Nanomaterials. 2019;**9**(2):164

[15] Shanmugam K, Browne C. Nanocellulose and its composite films: Applications, properties, fabrication methods, and their limitations. In: Nanoscale Processing. Elsevier; 2021. pp. 247-297

[16] Shanmugam K. Spray Coated Nanocellulose Films-Production, Characterisation and Applications [PhD thesis]. Monash University; 2019

[17] Varanasi S, Batchelor WJ. Rapid preparation of cellulose nanofibre sheet. Cellulose. 2013;**20**:211-215

[18] Shanmugam K, Doosthosseini H, Varanasi S, Garnier G, Batchelor W. Flexible spray coating process for smooth nanocellulose film production. Cellulose. 2018;**25**:1725-1741

[19] Beneventi D, Zeno E, Chaussy D. Rapid nanopaper production by spray deposition of concentrated microfibrillated cellulose slurries. Industrial Crops and Products. 2015;**72**: 200-205

[20] Shanmugam K, Varanasi S, Garnier G, Batchelor W. Rapid preparation of smooth nanocellulose films using spray coating. Cellulose. 2017;**24**:2669-2676

[21] Magnusson J. Method for Spraying of Free Standing 3D Structures with MFC: Creation and Development of a Method [Master thesis]. Faculty of Health Science and Technology, Department of Engineering and Chemical Science, Chemical Engineering, Karlstad University; June 14, 2016

[22] Beneventi D, Chaussy D, Curtil D, Zolin L, Gerbaldi C, Penazzi N. Highly porous paper loading with microfibrillated cellulose by spray coating on wet substrates. Industrial & Engineering Chemistry Research. 2014; **53**(27):10982-10989

[23] Alsaiari NS, Shanmugam K, Mothilal H, Ali D, Prabhu SV. Optimization of spraying process via response surface method for fabrication of cellulose nanofiber (CNF) film. Journal of Nanomaterials. 2022; **2022**:1-10

[24] Vikman M, Vartiainen J, Tsitko I, Korhonen P. Biodegradability and compostability of nanofibrillar cellulose-based products. Journal of Polymers and the Environment. 2015;**23**:206-215

[25] Shanmugam K, Chandrasekar N, Balaji R. Barrier performance of spray coated cellulose nanofibre film. In: Micro. MDPI; 2023. pp. 192-207

Section 2

Properties and Applications of Nanocellulose

Chapter 3

Nature-Inspired Nano Cellulose Materials, Advancements in Nano Cellulose Preparation and Versatile Applications

Hanumanthu Jeevan Rao, Sanjay Singh, Perumalla Janaki Ramulu, Narender Singh, Thiago F. Santos, Caroliny M. Santos, Nandini Robin Nadar and Gara Dheeraj Kumar

Abstract

A promising ecofriendly, cost-effective biomaterial derived from natural sources, i.e., lignocellulose natural fibers from trees, plants, agri waste, fruits, vegetables, seeds, and leaves. It clicked the attention of the researchers due to promising properties and versatility. The aim of the study is to examine the recent developments and preparation methods and extraction techniques for nanolignocellulose materials from nature. It also discusses the wide range of applications that use nano cellulose's remarkable properties for a variety of research fields. Current work discussed advancements in nano cellulose preparation techniques, innovative methods, and cutting-edge applications. The ease of nano cellulose excels as a material for tissue engineering scaffolds, wound dressings, flexible and sustainable electronics, and drug delivery systems in the biomedical industry. It is desirable component in composites due to its remarkable reinforcing abilities in polymers. The use of environmental applications such as water purification, oil spill cleanup, and biodegradable packaging is also highlighted in this research. The studies emphasize the need for more investigation and optimization of extraction processes, characterization, and applications. This multidisciplinary study intends to motivate academics and scientists to fully utilize nano cellulose and contribute to the creation of environmentally friendly and sustainable solutions across a range of industries.

Keywords: nano cellulose, nature-inspired materials, nano cellulose applications, natural fibers, nanocomposites, sustainable materials

1. Introduction

Nanocellulose has received significant interest due to its mechanical and optical properties [1], biodegradability [2], availability [3], recyclability [4], renewability [5], and low coefficient of thermal expansion (CTE) [6] as exhibited in **Figure 1**. This

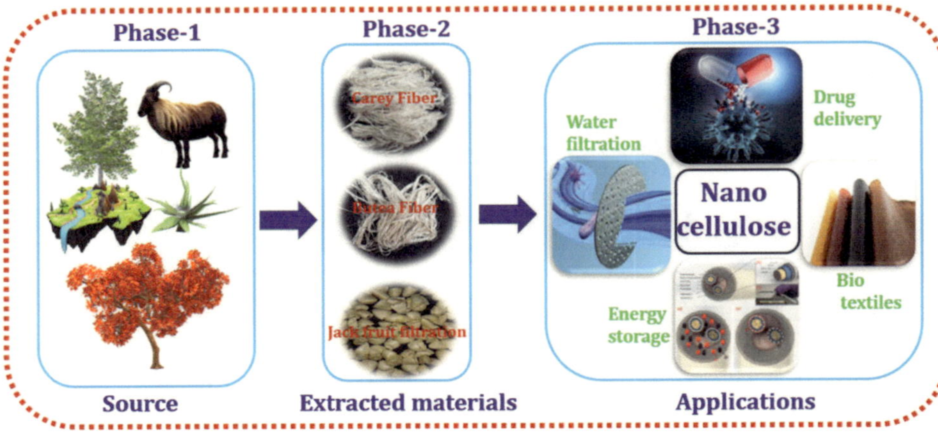

Figure 1.
Availability, extraction, and applications of nano cellulose.

nanomaterial has been reported as an intriguing sustainable and tunable platform for the manufacturing of a variety of high-value products. It can take the shape of nanofibrils or nanocrystals [7]. Microcellulose materials derived from nature constitute a promising class of biomaterials distinguished by remarkable qualities and a broad range of possible applications. These naturally occurring materials have attracted a lot of interest because of their exceptional mechanical strength, large surface area, biocompatibility, and environmentally favorable characteristics [8, 9]. An increasingly popular substitute for synthetic materials in a variety of industries is nanocellulose, a derivative made from abundant and sustainable plant sources [10]. With a focus on cutting-edge extraction methods, this study provides a thorough review of the most recent developments in nanocellulose synthesis and its many applications. Cellulose, the most common biopolymer on Earth, serves as the fundamental building element of plant cell walls. Microfibrils make up its hierarchical structure, which can be further broken down into nanoscale dimensions [11]. Numerous extraction techniques are used to produce nanocellulose, each of which has unique benefits and features. Techniques for mechanical disintegration that efficiently shrink cellulose fibers to nanoscale dimensions include high-pressure homogenization and ultrasonication [12]. Meanwhile, amorphous portions are selectively removed using chemical methods such acid hydrolysis and oxidation, producing extremely crystalline nanocellulose [13]. Furthermore, environmentally conscious extraction methods involve enzymatic processes that harness cellulase enzymes to derive nanocellulose from cellulose-rich sources [14].

In recent trends, the efficiency and sustainability of nanocellulose extraction processes have taken center stage. The quest for greener alternatives has led to the exploration of eco-friendly solvents, aiming to replace traditional chemicals and minimize environmental impact. Enzymatic techniques have gained prominence due to their exceptional selectivity, low energy requirements, and minimal waste generation. Researchers have also investigated synergistic combinations of multiple extraction methods to maximize nanocellulose yield and enhance its properties. The overarching objective of ongoing research endeavors is to establish scalable and cost-effective extraction methodologies that enable large-scale production of nanocellulose for industrial applications [15]. In essence, the remarkable attributes of nature-inspired nanocellulose materials, coupled with advancements in extraction techniques, hold the promise of revolutionizing various industries through sustainable and innovative solutions [16]. This article provides a comprehensive exploration of these breakthroughs, shedding light on the potential of nanocellulose as a versatile and environmentally conscious biomaterial as shown in **Figure 2**.

Its characteristics have been further improved and its prospective uses have been increased because of improvements in nanocellulose preparation procedures [17]. Functional additives have been added to nano cellulose to give it certain functions, enabling customized applications in a variety of industries. These additions include nanoparticles, polymers, and biomolecules. To fine-tune the surface features and increase interfacial adhesion in composite materials, surface modifications through grafting or chemical treatments have been used. These preparation methods have led the way for the creation of hydrogels, films, coatings, and composites based on nanocellulose that exhibit extraordinary properties and have a wide range of applications, from electronics to healthcare [18]. Nano cellulose's adaptability is demonstrated in a wide range of study fields. Due to their biocompatibility and biodegradability, nano cellulose-based materials are used in biomedicine as diagnostic platforms, scaffolds for tissue engineering, wound dressings, and drug delivery systems [19]. Taking use of its superior mechanical qualities and electrical conductivity, nano cellulose has demonstrated promise as a sustainable option for flexible electronic devices and sensors in nanoelectronics. Additionally, due to its remarkable adsorption capability and biodegradability, nano cellulose has attracted interest in environmental applications such as water filtration, oil spill repair, and sustainable packaging options. A paradigm change in materials research and engineering has been sparked by nature-inspired nano cellulose materials, which provide a green and sustainable alternative to traditional materials. The goal of this study work is to present a thorough grasp of the developments in nano cellulose preparation methods, flexible applications, and the most recent extraction method trends. It is believed that by putting light on the most recent advancements in this area, researchers and scientists would be motivated to investigate the enormous potential of nano cellulose, advancing the development of sustainable technologies and applications in a variety of industries.

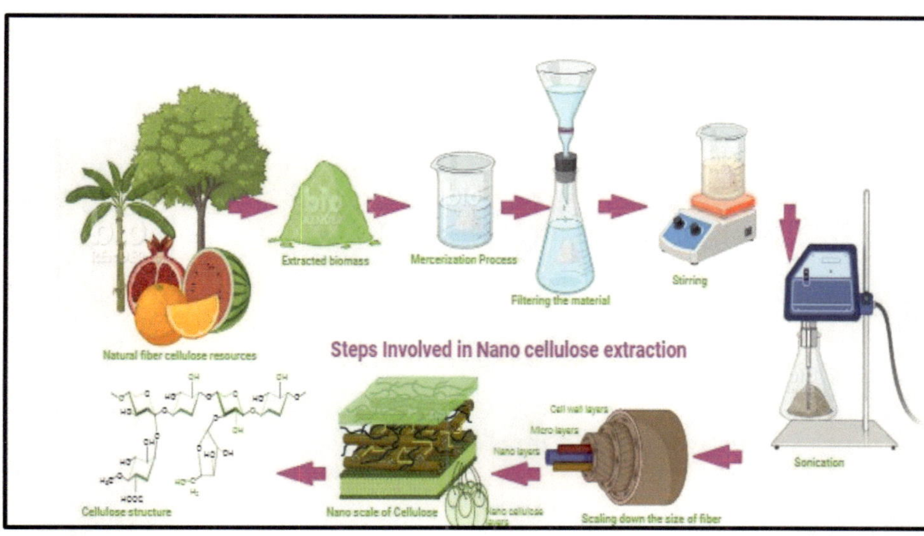

Figure 2.
Extraction process of nano cellulose.

2. Work carried by the researchers in the last decade

The number of papers published annually varied from 2012 to 2022, but it witnessed a rise as of 2017 with 38 articles and reached a peak in 2022 with 72 published articles. **Figure 3** depicts the annual trends in publications on this topic (nano cellulose; preparation; synthesis; applications) based on a sample of 436 articles gathered on August 2022. As shown in **Figure 3**, the publication on nanocellulose for versatile applications increases every year, especially for packaging, capacitors, batteries, drug delivery, water filtration, and sensing devices applications [20]. The rise in papers discussing the use of nanocellulose for batteries, packaging, capacitors, medicine delivery, water filtering, and sensing devices is a clear sign of the material's possible future uses. Because of its exceptional physiochemical and biological properties, nanocellulose has emerged as a material with potential for a variety of applications thanks to the rapid growth of nanotechnology and materials science. According to **Figure 3**, the number of articles has been increasing exponentially while the number of citations has decreased during the past 5 years in the field of nanocellulose. This was due to a balance between new discoveries and existing advances. This may result in a temporary decrease in citations as the scientific community works to incorporate and evaluate the newly available information. Considering the aforementioned properties of nanocellulose, transforming this renewable material, which reduces carbon footprint into advanced products is a crucial step toward sustainable development [21].

2.1 2010–2012

The researchers aimed to identify and develop the various natural fibers and micro cellulose and nano cellulose extraction techniques for lignocellulose materials. Various mechanical techniques such as high-pressure homogenization and ultrasonication were

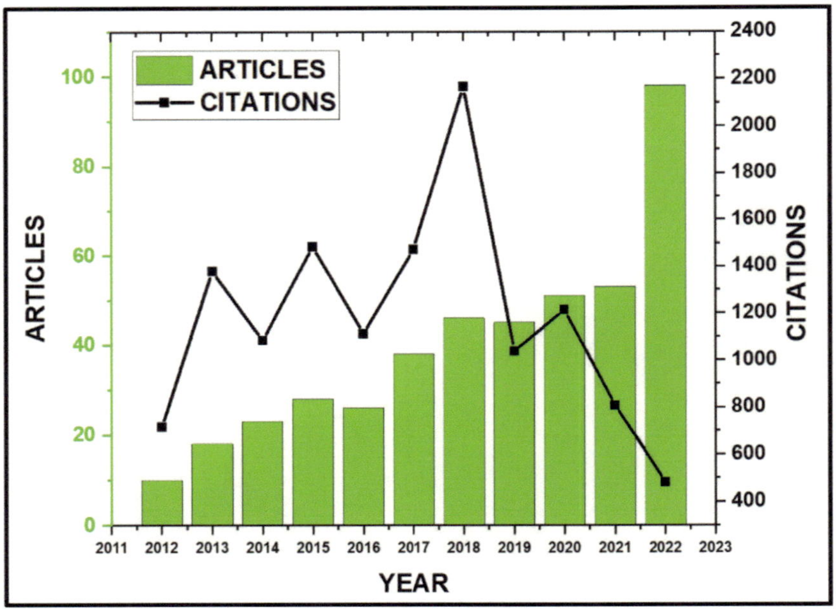

Figure 3.
Articles published annually from 2012 to 2022 in "nano AND cellulose; preparation* OR syntes*; applications".*

identified to extract the desired efficient micro and nano fibrils and, they break down lignocellulose biomass into nano scale [22, 23]. Different chemical treatments, such as acid hydrolysis and oxidative methods, were studied, implemented, and investigated to convert the material phases and generation of crystalline nano lignocellulose [24]. In addition to that, enzymatic approaches received attention due to sustainable and eco-friendly nature and particularly in producing cellulose to nano fibrils and particles.

2.2 2013–2015

Scientific community focused on multiplying the productivity and adaptability for suitable applications of nano cellulose development methods. New and eco-friendly solutions, chemical combinations were developed to substitute older chemicals in extraction methods, intended to minimize the environmental effects. Also, there is development in combination of different techniques to improve the quality of the materials extractions and these combinations minimize the energy consumption and time for the processes [25]. Also, researchers focused on characterization of the Nano lignocellulose and investigations carrying on its structural, mechanical, chemical, thermal and Morphological properties for suitable sustainable applications.

2.3 2016–2018

The enhancement in the nano cellulose excitation continued with various sources, also the functionalization of the materials is very important in the applications like energy storage, water filtration, drug delivery and bio textiles during the extraction process [26, 27]. Also, the scientific community started adding some nano materials to improve the specific properties nano cellulose, making it more adaptive for desired applications. By using various chemical treatments, the surface modifications of nano cellulose are improved and investigated various combinations, it is enhancing the different parameter of the nano cellulose thermal stability, crystallinity, surface roughness, morphology is very useful for development of advanced composite materials [28].

2.4 2019–2022

Researchers are currently concentrating more on the difficulties associated with extracting a significant amount of nanocellulose from different types of biomasses [29–31] and their byproducts as shown in **Figure 4**. Numerous adjustments were made to extraction procedures to improve production effectiveness and lower costs, making nanocellulose economically feasible for a variety of sectors. To fulfill industrial objectives, scaling-up enzymatic procedures and continuous flow extraction techniques became popular. To further increase the accessibility and sustainability of nanocellulose, researchers looked into other sources of cellulose-rich materials [32–37]. Interdisciplinary research partnerships also developed, bringing together specialists from diverse sectors to create cutting-edge, environmentally friendly extraction methods. Research efforts to comprehend the potential negative effects of nanocellulose extraction on human health and the environment increased, resulting in the creation of safer and more environmentally friendly extraction techniques. Overall, the field of nano cellulose extraction made tremendous strides between 2010 and 2022 as researchers worked to increase the effectiveness, scalability, and sustainability of extraction techniques. These developments pave the way for the widespread use of nanocellulose materials inspired by nature in a variety of fields, such as biomedicine, electronics, packaging, and environmental remediation [38].

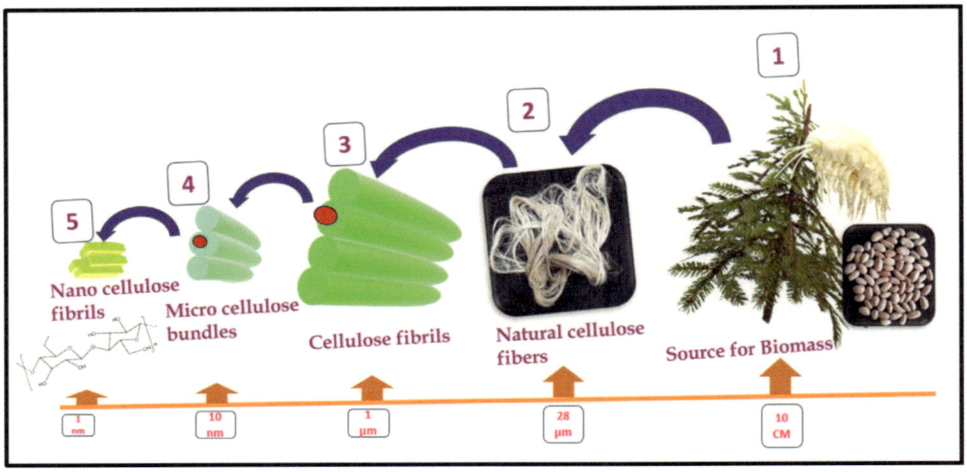

Figure 4.
Anotomy and stages of Nano cellulose.

3. Materials and methods

3.1 Fourier transform infrared (FTIR) analysis

FTIR identified characteristic peaks for cellulose, hemicellulose, and lignin in the biomass-based nano cellulose was carried out using a Thermos Scientific Nicolet iS50 FTIR Spectrometer, and the characterization revealed the real functional groups connected to the with nano cellulose sample. Crystallinity degree of nano cellulose was determined, indicating its potential for structural applications, FTIR analysis provided insights into the thermal stability of the nano cellulose, relevant for its processing and applications to view the FTIR spectra on any nano cellulose particles, it mixed with potassium bromide in the appropriate ratio, and samples are subsequently pelletized using an XLR pellet press. With a scan rate resolution, the FTIR spectra identified the range between 4000 and 400 cm^{-1}.

3.2 X-ray diffraction (XRD) analysis

XRD helps quantify the crystallinity index (CI), a measure of the ordered crystalline portion of the nano cellulose. High CI indicates strong cellulosic chains and potentially superior mechanical properties. XRD peak broadening can be used to estimate the average size of crystalline domains (crystallites) in the nano cellulose. Larger crystallites often improve strength and stiffness, while smaller ones enhance surface area and reactivity. XRD patterns can also provide information about crystal morphology, such as the presence of preferred orientations. Besides cellulose, natural fibers may contain other components like hemicellulose, lignin, and minerals. XRD can identify these impurities by their characteristic peaks, allowing assessment of the purity of the extracted nano cellulose. This knowledge is crucial for understanding the properties and potential applications of the material. Different XRD peak intensities and profiles reveal the arrangement of cellulose chains within the orientations. The Bruker D8 Advance X-Ray diffraction instrument was used to compute the nano cellulose's crystallography and identify its crystalline characteristics. The mechanical, thermal, and tribological properties of the fiber were affected by these crystalline effects on a small number of the nanocellulose particles that were characterized. With a spectrum magnification mode of 2 and a step size of 0.02° at room temperature, the instrument scanning range ranged from 10° to 80°.

3.3 Thermogravimetry analysis (TGA)

TGA measures the mass loss of a material as its temperature increases. For nano cellulose, TGA identifies the decomposition temperature, marking the onset of significant mass loss. A higher decomposition temperature indicates greater thermal stability, making the material suitable for high-temperature applications. TGA also reveals the onset temperature, at which initial mass loss starts, providing insights into the initial thermal behavior. TGA curves help elucidate the mechanisms of thermal degradation in nano cellulose. Multiple mass loss steps often indicate the breakdown of different components like cellulose, hemicellulose, and lignin at different temperatures. Analyzing the weight loss percentages and temperature ranges can reveal the dominant degradation mechanisms, such as dehydration, depolymerization, and charring, TGA can be used to assess the impact of processing methods and chemical modifications on the thermal stability of nano cellulose. Treatments like mercerization or acetylation can alter the thermal degradation profile by influencing crystallinity, chain structure, and functional groups. Understanding these effects helps optimize processing and tailor nano cellulose for specific thermal performance requirements Thermo-gravimetric analysis (TGA), which was carried out with the TGA-DTA Perkin Elmer STA6000, Perkin Elmer Diamond, and an instrument with an integrated Pure Platinum Pan Furnace Holder with Ring for Sample Handling with Temperature Rise and Appropriate Balancing, revealed the separation of functional groups of the nano cellulose regarding weight loss. The loaded fibril specimen's thermal conditions steadily increase up to 20 C to 1000 C with a temperature ramp of 20.00 C/min. However, pure Nitrogen is delivered at a 20 ml/min effective flow rate, maintaining the camber conditions.

3.4 Differential scanning calorimetry (DSC)

By using a Netzsch DSC 204f1 Phoenix Instruments, researchers were able to measure the thermal characteristics of nanocellulose in relation to the material's transition temperature as well as the amount of heat emitted by the sample. The thermal transition rate was 20–400°C. In conjunction with nitrogen gas, the sample was placed in an aluminum tray with a pierced cap, which escalated the temperature until the fiber took the form of ash.

3.5 Transmission electron microscopy (TEM)

TEM offers high-resolution images of NFC, revealing its individual fibrils, their diameter, length, and surface features. This level of detail is crucial for understanding the influence of processing methods (e.g., acid hydrolysis, mechanical fibrillation) on the morphology of NFC, which impacts its properties and applications. TEM analysis can be coupled with selected area electron diffraction (SAED) to examine the crystal structure of NFC.SAED patterns reveal the arrangement of cellulose chains within the crystallites, providing information about the crystallinity index (CI) and crystal orientation. High CI indicates a well-ordered crystalline structure, potentially leading to superior mechanical properties in NFC-based materials. TEM can also detect the presence of aggregates, bundles, or entangled regions within the NFC structure, providing insights into its dispersion behavior. Transmission electron microscopy is typically employed by Meditec, Inc., Oberkochen, Germany, to determine the micro anatomy of nanocellulose. Prior to bath sonication, samples were prepared as 200–300 nm-sized particles in water and ethanol. To hasten the volatilization of fluids from the grid, ethanol was added. Also, TEM can be used to investigate specific features of NFC, such as the presence of pores, internal defects, or surface modifications by chemical treatments. It allows visualization of functional groups attached to the surface, which are crucial for further modifications and interactions with other materials. This information is valuable for tailoring NFC for specific applications like composites, biofilms, or drug delivery systems.

3.6 Scanning electron microscopy (SEM)

SEM offers high-magnification images, revealing the surface texture, roughness, and presence of features like pores, cracks, or fibril bundles in NFC.

This information is crucial for understanding the adhesion and interaction of NFC with other materials, impacting its performance in composites, membranes, or bio-based materials. SEM can also visualize the distribution of chemical modifications or coatings applied to the NFC surface, providing insights into its surface functionality. SEM allows observation of the arrangement and connectivity of individual NFC fibrils within the network. It can reveal the presence of entanglement, aggregation, or bridging between fibrils, influencing the overall porosity, strength, and stability of the NFC scaffold. Understanding the network structure helps tailor processing methods to achieve desired properties in NFC-based materials like gels, films, or scaffolds for tissue engineering. SEM can identify and characterize the presence of any remaining impurities or contaminants in the NFC, such as residual lignin, hemicellulose, or mineral particles. These impurities can affect the mechanical properties, surface chemistry, and functionality of NFC and therefore need to be minimized for specific applications. By providing visual evidence of their presence and distribution, SEM guides optimization of purification processes for obtaining high-quality NFCWith the JEOL - JSM-6390 - Scanning Electron Microscope (SEM) instrument, the scanning of the Nano lignocellulose is observed with a focused beam of electrons. The electrons interact with Nano lignocellulose atoms to produce the topography of the fiber surface with quality images seen in high convention scanning electron microscopy mode with 4.0 nm, and the electron beam's accelerating voltage is 0.5 to 30 kV at room temperature.

3.7 Particle size analyzer

A particle size analyzer (Nano-ZS, Malvern Instruments Ltd., UK) was used to measure the physical parameters of nanocellulose, including diameter, size distribution, and zeta potential. Just before analysis, the samples were diluted by deionized water five times to measure zeta potential.

3.8 Atomic force microscopy (AFM)

Unlike SEM, AFM operates at the atomic level, providing incredibly detailed images of the NFC surface. It can visualize individual cellulose chains, their orientation, and surface features like pores, roughness, and defects with unparalleled resolution. This information is crucial for understanding the surface chemistry, reactivity, and interaction of NFC with other materials at the molecular level. AFM goes beyond imaging by measuring the surface forces and mechanical properties of NFC at the nanoscale. It can determine the elasticity, stiffness, adhesion, and friction of individual fibrils within the NFC network. This knowledge contributes to tailoring NFC for specific applications requiring high strength, flexibility, or controlled interactions with other materials. AFM can be used to directly observe the distribution and morphology of chemical modifications or coatings applied to the NFC surface. It can visualize grafted functional groups, polymer attachments, or nanoparticles deposited on the cellulose chains. This information is vital for understanding the effectiveness of surface modifications in enhancing specific properties of NFC for applications like filtration, drug delivery, or biocatalysis. The potential information of Nano cellulose surface roughness, average surface roughness-(Ra), Root mean square (RMS)roughness-(Rq), Different point average surface roughness-(Rz), Skewness-(Rsk), Kurtosis-(RKU), and it exhibits the highest peak-to-valley height (Rt), was observed using the scanning probe microscopy of AFM technique. Asylum Research, MFP-3D BIO AFM equipment (USA) was used to perform the scanning.

3.9 Energy dispersive X-ray analysis (EDAX)

To determine the primary elements and chemical combinations on the surface of the nanocellulose particles, this analytical technique focuses on elemental analysis using X-ray excitation. The summary of several compounds and elements, including Ni, Be, Ti, Zn, C, O, Na, Mg, Al, Si, Ag, P, Cl, K, Ca, and S, is distributed. The model Jeol 6390LA scanning electron microscope was characterized (**Table 1**).

Forest residues	Extraction technique	Properties	Reference
A. mangium (wood)	Pulping: 17% active alkali, 170°C, 3.5 h. Bleaching: DEDED (chlorine dioxide-alkali extraction-chlorine dioxide-alkali extraction-chlorine dioxide). Hydrolysis: 64% (w/w) H_2SO_4, 45°C, 40 min. Dialysis and 25 min sonication.	Length range: 152.6 to 238.4 nm, width range: 5.65 to 10.70 nm. Crystallinity: 79%	[39]
Apple tree prunings	Pre-treatment: 7.5% (w/w) NaOH, 90 min and 60% (v/v) acetic acid, 180°C, 90 min. Bleaching: 2% (w/w) NaOH, room temperature overnight. Acetate buffer and 1.7% (w/w) of $NaClO_2$, 70°C, 2 h. Acid hydrolysis: 64% (w/w) H_2SO_4, 45°C, 45 min. Dialysis 7 days, ultrasonication 3 min.	Smaller and more homogeneous size (aprx. 300 nm and PDI < 0.2) High crystallinity index: 84%.	[40]
Bamboo fiber	Alkaline Pre-treatment: 17% (w/v) NaOH, 170°C, 90 min. Bleaching: NaClO and acetic acid, 70–80°C, 30 min. NaOH, 30 min. Acid hydrolysis: 64% (w/w) H_2SO_4, 45°C for 45 min.	Higher degree of crystallinity and higher yield of CNCs about 86.96% and 22% respectively	[41]

Forest residues	Extraction technique	Properties	Reference
Bamboo fibers	Kraft pulping: 20% active alkali and 30% sulphide (ratio 1:7), 170°C, 3 h, 12–14 bar pressure. Bleaching: NaClO, 70–80°C, pH 12, 1 h, and acetic acid. 8% (w/v) NaOH, room temperature, 30 min. Acid hydrolysis: 2.5 mol/L HCl, 85°C, 30 min.	The peak temperature: 353°C High crystallinity index of 82.6%. Greater thermal stability	[41]
Bamboo log chips	Glycerol pretreatment, extrusion treatment and mechanical refining using 0.15% sulfuric acid (conc.) as a catalyst.	Diameter: 20–80 nm, Crystallinity: 52.7%.	
Beech wastes pulp	Fibrilization	Diameters: 20–65 nm, Average: 35 nm.	[42]
Betung bamboo	Isolation of α-cellulose: 3.5% (v/v) HNO_3, 90°C, 2 h. NaOH and Na_2SO_3, 2% (w/v) each, 50°C, 1 h. Water and 3.5% (w/v) NaClO. 17.5% (w/v) NaOH, 80°C, 30 min. Acid hydrolysis: 2.5 N HCl, 15 min.	Particle size distribution 1117.4 nm, pH 6.88, ash contents ±0.0584%, moisture content 36%, loss on drying 4.59%.	[43]
Birch and Spruce sawdust	Sodium hydroxide and Soxhlet extraction followed by acetic acid, sodium acetate and sodium chlorite treatment	Tensile strength:80–200 MPa; Young's modulus: 4.8–8.5 GPa.	[44]
Bleached hardwood pulp	Chemical method: 50–85% (v/v) phosphotungstic acid, 90°C, 15–30 h and diethyl ether.	Particles size: 15–40 nm. Good thermal stability	[45]
Bleached Kraft bamboo pulp	Acid hydrolysis: H_2SO_4 0.8% (v/v) HCl, 165°C, 40 min.	Narrow particle diameter: 30–70 μm. High crystallinity: 78.7%	[46]
Cordia goeldiana veneer wastes	Alkali treatment, bleaching, homogenization, and casting.	Maximum processing temperatures: 300 C.	[47]
Calotropis procera fiber	Pretreatment: 2% (w/v) NaOH, 3 h. 93% (v/v) acetic acid and 0.3% (v/v) HCL, 90°C, 3 h. Bleaching: 5% (v/v) H_2O_2 and 3.8% (w/v) NaOH, 3 h. Acid hydrolysis: 63% (w/w) H_2SO_4, room temperature, 1 h. Dialysis 2 days and ultrasonication.	Average diameter: 12 nm Average length: 250 nm. Aspect ratio of approximately 30. Crystallinity index: 68.7%. Good thermal stability	[48]
Date seeds	Dewaxing: chloroform and ethanol (2:1). Delignification: 17.5% (w/v) NaOH, 90°C, 3 h. Bleaching: NaClO, 80°C, 45 min. Acid hydrolysis: 2.5 N HCl, (105 ± 2)°C, 45 min.	Particle Size: 100–300 μm. Thermal stability increased. Crystallinity: 70%	[49]
Eucalyptus sawdust	TEMPO oxidation	Surface area: 60 $m^2 g^1$; Average diameter: 41.0 nm.	[50]
Kans grass	Alkaline treatment: 3% (w/w) NaOH, 14 h, autoclaving at 137,895 Pa and 210 ± 5°C, 45 min. Bleaching: H2O2 and NaOH.	The highest yield: 83% Crystallinity: 74.06% Peak temperature: 338°C,	[51]
Lemang and Semantan bamboo	Pulping: 23% (w/v) NaOH and 0.1% anthraquinone, 160°C, 2 h. Alkaline treatment: 17.5% (w/v) NaOH, 80°C. Acid hydrolysis: 2.5 N HCl, 100°C, 30 min.	Improvement in Tensile Strength: 5.75%, Improved Elongation Value: 5%	[52]
Logging residues	Alkaline treatment, bleaching treatment, and acid hydrolysis	High aspect ratio:>10; Good thermal stability.	[53]

Forest residues	Extraction technique	Properties	Reference
Medium density fiberboard	Soxhlet extraction, sodium hydroxide and repeated bleaching.	Length: 164.7 nm; Width: 6.7 nm; Crystallinity: 71%,	[54]
Oil palm empty fruit bunch	Pretreatment: 4% (w/v) NaOH, 80°C, 2 h and dimethyl sulfoxide, 80°C. Acid hydrolysis: 64% (w/v) H_2SO_4, 45°C, 2 h. Dialysis 3 days, ultrasonication 30 min.	High cellulose content: 87.7%. Higher Crystallinity: 80%. Higher thermal stability	[55]
Oil palm empty fruit bunch	Preparation of MCC: Oxygen, ozone, peroxide for bleaching and 2.5 M HCl, (105 ± 2)°C, 15 min for acid hydrolysis. Preparation of NCC: 4-acetamido-TEMPO and sodium bromide, 10% (w/v) NaClO, 30°C for 2 h. Sonication 30 min.	High dispersion stability with yields of 93%, Decomposition temperatures: 200°C. Average length: 122 Average width: 6 nm, Crystallinity: 72%	[56]
Pine needles	Chemical pretreatments followed by ultrasonic treatments	Narrow diameter: 30–70 nm; Cellulose I type; Crystallinity: 66.19%; Highly flexible, highly ultralight, and good thermal properties	[57]
Pinecone	Alkaline treatment: 4.5% (w/v) NaOH, 80°C, 2 h. Bleaching: mixture of 1.7% (w/v) aqueous NaClO2, 0.2 M acetate buffer and water, 80°C, 4 h. Acid hydrolysis: 65% (w/w) H2SO4, 45°C, 30-, 45-, and 90-min. Dialysis 1 week, ultrasonication 5 min.	Average diameter < 3 mm Average length < 335 nm High aspect ratio, the high crystallinity, and the good thermal stability	[58]
Pinecone biomass	Acidification, alkali treatment, mechanical grinding.	Tensile strength: 273 MPa; Elastic modulus: 17 GPa; Crystallinity: 70%. Diameter: 5–20 nm	[59]
Rubber-wood and kenaf-bast fibers	Alkaline treatment: 2 M NaOH, room temperature, overnight. Bleaching: 35% (v/v) H_2O_2, 75°C, 4 h. Acid hydrolysis: 48% (w/w) H_2SO_4, 45°C, 120 min. Dialysis 3 days.	Average diameter 5.14 nm for Rubber and 5.27 nm for kenaf. Crystallinity index of Rubber and kenaf was 74.34% and 73.19% respectively.	[60]
Sacred Bali bamboo	Pulping: 23% (w/v) NaOH and 0.1% anthraquinone, 160°C, 2 h. Bleaching: $NaClO_2$ and 10% (v/v) acetic acid, 85°C, 2 h. 17.5% (w/v) NaOH, 80°C, 1 h. Acid hydrolysis: 2.5 N HCl, 100°C, 30 min.	Higher Tensile strength Fiber length: 0.47 mm Crystallinity index (%):78.00	[61]
Softwood forest residues [woody chips, branches, and pine needles]	Alkaline treatment: 4.5% (w/v) NaOH, 80°C, 2 h. Bleaching treatment: Mixture of 2 M acetate buffer, 1.7% (w/v) aqueous chlorite and water, 80°C, 4 h. Acid hydrolysis: 65% (w/w) H_2SO_4, 45°C, 40 min. Dialysis 1 week and ultrasonication 10 min.	Gravimetric yield of over 13%. Higher crystallinity Higher thermal stability	[62]

Table 1.
Nano cellulose source and extraction techniques.

4. Versatile applications of nature-inspired nano cellulose

While nature-inspired nano cellulose materials hold great promise in these applications, challenges remain in terms of scalability, cost-effectiveness, and standardization of production methods. Additionally, ensuring compatibility with existing recycling and composting systems is crucial for widespread adoption. The integration of nature-inspired nano cellulose materials in barrier coatings and biodegradable packaging solutions presents a transformative pathway for the packaging industry [62, 63]. These applications not only enhance product preservation and consumer experience but also contribute to a more sustainable and environmentally conscious future by reducing plastic waste and promoting circular economy principles. Continued research and development in this field are essential to fully unlock the potential of nature-inspired nano cellulose in addressing the evolving needs of modern packaging, innovative capacitors, and batteries, advanced drug delivery systems, efficient water filtration materials, and new sensing devices as showed in **Figure 5** [64, 65].

4.1 Barrier coatings and biodegradable packaging

Nature-inspired nano cellulose materials have demonstrated exceptional potential in revolutionizing packaging technologies by offering both barrier coatings for extending shelf life and innovative solutions for biodegradable and compostable packaging. These applications address critical challenges in the packaging industry, such as food preservation, environmental sustainability, and reduced plastic waste.

4.1.1 Barrier coatings to enhance packaging shelf life

A natural contender for extending the shelf life of many packaged items is nano cellulose, which has outstanding barrier characteristics against gases, moisture, and

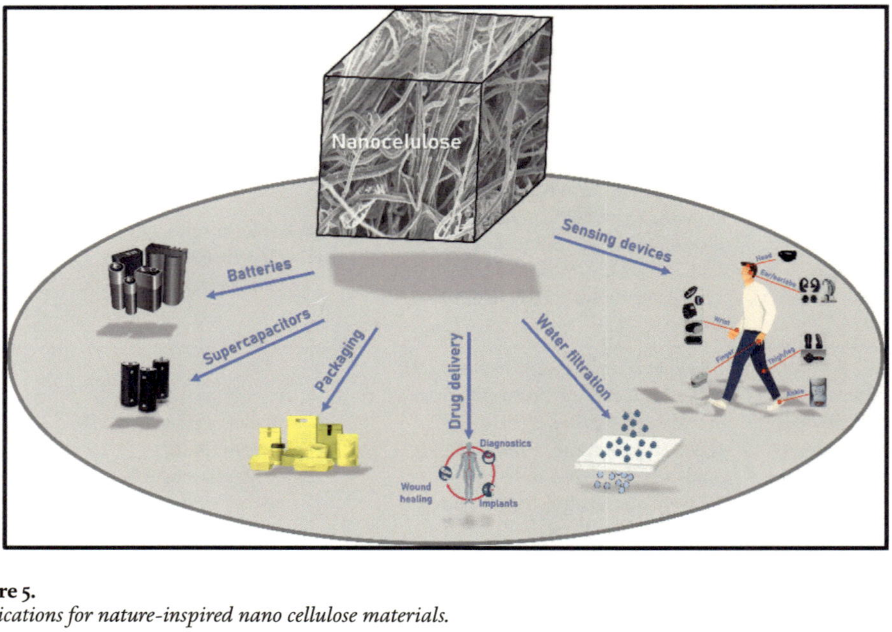

Figure 5.
Applications for nature-inspired nano cellulose materials.

DOI: http://dx.doi.org/10.5772/intechopen.114222

other external pollutants. Food waste can be decreased, and sustainable consumption encouraged by using nano cellulose as a barrier layer to dramatically extend the freshness and quality of perishable goods.

4.1.1.1 Benefits

- Improved gas and moisture barrier properties, minimizing the entry of oxygen and water vapor.
- Enhanced protection against flavor and aroma loss, maintaining product quality.
- extended shelf life due to increased resistance to external pollutants.
- fewer synthetic barrier materials are used, enhancing environmental sustainability.

4.1.1.2 Applications

- Fresh produce packaging (fruits, vegetables, herbs) to prevent spoilage and dehydration.
- Dairy and meat products packaging to preserve freshness and prevent bacterial growth.
- Snack foods and confectionery packaging to maintain crunchiness and flavor.

Lightweight, strength and mimicking nature's strength, nature inspired nano cellulose provides high strength-to-weight ratio, ideal for aircraft parts and bio-based composites. Derived from biomass like wood or plant waste, it promotes a circular economy and reduces environmental impact. It is safe for use in medical applications like wound dressings and drug delivery systems. Blocks moisture, oil, and gases, making it valuable for food packaging and protective coatings. Versatile modification allows tuning for specific applications like filtration, biosensing, and catalytic supports. Scalable production from various sources ensures cost-effectiveness and widespread accessibility. Enhances energy efficiency in buildings and materials due to its low thermal conductivity. Certain compositions offer inherent flame resistance, improving safety in construction and transportation. Adjustable pore size facilitates selective filtration, purification, and controlled release applications. Also, Paves the way for replacing petroleum-based materials with sustainable and eco-friendly alternatives.

4.1.2 Biodegradable and compostable packaging solutions

Biodegradable and compostable packaging options have drawn a lot of attention in response to the growing worries about plastic pollution and its effects on the environment. Given that it can be produced from renewable resources, exhibits biodegradability, and has a smaller environmental impact than traditional plastic packaging materials, nature-inspired nano cellulose presents an attractive option.

4.1.2.1 Benefits

- Renewable and sustainable material sources, reducing reliance on fossil fuels.
- Biodegradable and compostable, minimizing long-term environmental impact.
- Reducing marine pollution and plastic waste.
- Possibility of individualized mechanical characteristics and structural layouts.

4.1.2.2 Applications

- Cutlery and utensils for one-time use that break down after being used.
- Disposable cups and takeaway containers for food and beverages.
- Containers for bakery goods, snacks, and dry goods.
- Personal care product packaging, such as shampoo and soap.

4.2 Functional nanocomposites applications of nature-inspired nano cellulose

Nature-inspired nano cellulose has emerged as a remarkable candidate for enhancing the performance of polymer composites, offering a wide array of mechanical, thermal, and electrical enhancements [66]. The integration of nano cellulose into polymer matrices imparts unique properties that open up innovative applications in various industries. This section explores the multifaceted applications of nano cellulose as a reinforcement in polymer composites, highlighting its contributions to mechanical strength, thermal stability, and electrical conductivity [67, 68]. While significant advancements have been made in the development of nano cellulose-based functional nanocomposites, ongoing research aims to further optimize dispersion techniques, tailor surface modifications, and explore novel polymer/nano cellulose combinations. These efforts will likely lead to even more diverse and advanced applications in fields such as smart materials, 3D printing, and multifunctional coatings [69]. The incorporation of nature-inspired nano cellulose into polymer composites for mechanical, thermal, and electrical enhancements holds immense promise for a wide range of applications. By harnessing the remarkable properties of nano cellulose, researchers and industries are driving innovation toward sustainable and high-performance materials that address contemporary challenges and pave the way for a more efficient and environmentally conscious future.

4.2.1 Nano cellulose reinforcement in polymer composites

Nano cellulose, due to its exceptional mechanical properties and abundant surface functionalities, has gained attention as an effective reinforcement material in polymer composites. By dispersing nano cellulose within a polymer matrix, these composites exhibit enhanced mechanical strength, stiffness, and durability compared to their neat polymer counterparts [70]. Nano cellulose's complex network structure has a synergistic effect that distributes and absorbs stress, enhancing the composite's overall mechanical performance [71].

DOI: http://dx.doi.org/10.5772/intechopen.114222

4.2.2 Mechanical enhancements

The mechanical characteristics of polymer matrices are significantly enhanced by adding nanocellulose. Enhancing the strength and toughness of composite materials are nanocellulose fibers, which have a high aspect ratio and exceptional tensile strength [72, 73]. This reinforcement is especially useful in sectors that demand materials that are low weight but nevertheless durable, such as athletic products, aircraft constructions, and automotive components [74–76].

4.2.3 Thermal enhancements

Nano cellulose's low thermal expansion coefficient and high thermal stability make it a valuable component for enhancing the thermal performance of polymer composites. By integrating nanocellulose, it is possible to lessen the thermal expansion mismatch between the polymer and other materials, potentially reducing warping or cracking due to temperature changes. This is helpful for applications like electronic packaging, thermal insulation, and flame-resistant materials where thermal stability is important.

4.2.4 Electrical enhancements

Nano cellulose can be used into polymer composites to add electrical conductivity because of its special electrical characteristics, including its high surface area and capacity to carry charges. By adjusting the nanocellulose quantity and dispersion within the polymer matrix, this conductivity can be adjusted. These conductive composites are used in wearable technology, flexible electronics, sensors, and electromagnetic interference shielding.

4.2.5 Environmental benefits

The use of nature-inspired nano cellulose as a reinforcement in polymer composites aligns with sustainability goals. Nano cellulose is derived from renewable sources and is biodegradable, making it an eco-friendly alternative to synthetic reinforcements. The reduced reliance on fossil-based materials contributes to a more environmentally responsible approach to composite manufacturing.

4.3 Biomedical applications of nature-inspired nano cellulose

Nanocellulose materials that are inspired by nature have shown incredible promise in the field of biomedicine, providing ground-breaking approaches to drug delivery, wound healing, and tissue engineering [77]. Nanocellulose-based materials have distinctive features that make them ideal candidates for many biomedical applications, drawing inspiration from the intricate architectures present in natural systems [78]. There are obstacles to overcome despite the apparent potential of nature-inspired nanocellulose materials in biomedical applications. In order to move these materials from the laboratory to clinical practice, it is essential to ensure batch-to-batch consistency, maximize scalability, and obtain regulatory approvals. To ensure patient safety, a thorough assessment of the long-term biocompatibility and degrading behavior of nanocellulose-based materials is required [79]. Nano cellulose materials with natural inspirations have opened up new avenues for biomedical uses. Nano cellulose-based materials provide inventive solutions that take advantage of their special qualities and biomimetic traits, including adaptable drug delivery carriers, wound healing assistance, and tissue engineering [80]. These substances have the potential to revolutionize how we approach healthcare and regenerative medicine as research develops, leading to better patient outcomes and quality of life.

4.3.1 Drug delivery carriers

Materials made of nanocellulose have shown promise as regulated and precise medication delivery systems. Their large surface area, adaptable surface chemistry, and biocompatibility make it possible for medicinal substances to be loaded and released quickly. Numerous medications, including hydrophilic and hydrophobic substances, proteins, and nucleic acids, can be enclosed in nanocellulose. Nano cellulose's porosity structure permits continuous and regulated medication release, reducing adverse effects and improving therapeutic results. Precision medicine techniques are made possible by the surface modification of nanocellulose with targeting ligands, which increases its specificity to cell types or regions.

4.3.2 Wound healing materials

Wound dressings made of nano cellulose have drawn attention because of their outstanding abilities to speed up the healing process. These substances have a great capacity to store water and can keep a wound wet, which promotes cell migration, proliferation, and tissue regeneration. Nano cellulose dressings also possess antibacterial qualities, which can aid in the healing process and help prevent infections. Nanocellulose wound dressings resemble the extracellular matrix of tissues to give the wound mechanical support and can be customized for various wound types and stages.

4.3.3 Tissue engineering scaffolds for regenerative medicine

Nano cellulose scaffolds are essential for tissue engineering because they provide a three-dimensional environment for cell adhesion, growth, and differentiation. These scaffolds offer a favorable environment for tissue regeneration because they closely mirror the structural design of tissues and organs. The mechanical properties of nano cellulose can be modified to correspond to those of particular tissues, ensuring optimal mechanical support throughout the healing process. Nanocellulose scaffolds are excellent candidates for applications like bone, cartilage, and nerve tissue engineering because they can direct cell behavior and encourage tissue creation through surface functionalization and the insertion of bioactive chemicals.

4.4 Water purification and filtration of nature-inspired nano cellulose

Nature-inspired nano cellulose materials have shown promise in tackling water filtration and purification issues [81]. They are successful in eliminating impurities and raising the quality of the water due to their special characteristics.

4.4.1 Adsorbent materials for heavy metals and pollutant removal

Nano cellulose-based adsorbents are ideal for removing heavy metals and other contaminants from water sources because they have a large surface area and a lot of binding sites. Lead, mercury, cadmium, and arsenic can be effectively removed thanks to their natural affinity for metal ions. This application helps to protect public health and lessen the negative effects of industrial activity on the environment.

4.4.2 Microfiltration and ultrafiltration membranes

To precisely separate particles, bacteria, and macromolecules from water, nanocellulose materials can be designed into microfiltration and ultrafiltration membranes. These membranes can be functionalized to increase selectivity and permeability and have adjustable pore diameters. Their use in water treatment processes aids in the removal of suspended solids, bacteria, viruses, and other impurities, yielding cleaner and safer water resources.

4.4.3 Enhanced adsorption and filtration performance

The hierarchical structure of nano cellulose materials, reminiscent of natural systems, allows for tailored modification and functionalization. This flexibility enables the design of materials with enhanced adsorption capacities and filtration efficiencies. By optimizing surface chemistry and morphology, researchers can fine-tune the materials' performance for specific water treatment needs.

4.4.4 Sustainable water management

Incorporating nano cellulose-based materials into water purification and filtration technologies aligns with sustainability goals. These materials are derived from renewable resources and can be produced using eco-friendly processes, reducing the overall environmental impact of water treatment systems. The application of nature-inspired nano cellulose materials in water purification and filtration highlights their potential to contribute to cleaner and safer water sources. As ongoing research continues to advance these materials, their role in addressing global water challenges becomes increasingly significant. Whether through adsorption, membrane filtration, or tailored modifications, nano cellulose materials offer innovative solutions for sustainable water management and environmental protection [82]. Materials made of nanocellulose that are inspired by nature are useful for environmental cleanup following disasters or industrial accidents that taint water sources. Their effective adsorption and filtering abilities facilitate the quick removal of contaminants, assisting in ecosystem restoration and aquatic life preservation.

4.5 Sustainable energy solutions of nature-inspired nano cellulose

The performance of flexible electronics, energy storage devices, batteries, and capacitors has been significantly improved by nature-inspired nano cellulose, which has emerged as a promising contender for different sustainable energy solutions [83]. Nano cellulose's special qualities, such as its large surface area, mechanical toughness, and ion conductivity, make it a desirable material for developing energy innovations [84].

4.5.1 Nano cellulose in flexible electronics

- Materials that are not only lightweight and flexible but also ecologically benign are needed for flexible and wearable electronics. Nano cellulose, which is made from renewable resources, completely satisfies these requirements.

- The creation of wearable health monitoring devices, smart textiles, and flexible displays is made possible by the incorporation of nanocellulose-based conductive inks and films into flexible circuitry and sensors.

- Its biocompatibility and non-toxicity also make it appropriate for sensors and other implantable medical technology.

4.5.2 Energy storage devices

- Nano cellulose's porosity and large surface area provide an excellent platform for designing high-performance energy storage devices such as supercapacitors.

- Nano cellulose-based electrodes exhibit improved charge storage capabilities, enhancing energy density and power density of supercapacitors.

- Its inherent ability to absorb and retain electrolytes contributes to the overall performance and stability of the devices.

4.5.3 Enhancing battery performance

- Incorporating nano cellulose into battery electrodes can significantly enhance their performance by improving ion diffusion and electron transport.

- Nano cellulose's three-dimensional network structure can facilitate efficient ion and electron transfer, leading to faster charging and discharging rates.

- Enhanced conductivity and mechanical stability can contribute to prolonged cycle life and overall battery longevity.

4.5.4 Capacitors with nano cellulose

- Nano cellulose-based dielectric materials can be utilized in capacitors to enhance energy storage efficiency and stability.

- These materials can increase the energy density of capacitors while maintaining low levels of energy loss (dielectric loss).

- The eco-friendly and sustainable nature of nano cellulose aligns well with the growing demand for green energy solutions.

- The incorporation of nanocellulose materials derived from nature into sustainable energy solutions highlights the significance of materials science in addressing the need for cleaner and more effective energy technologies on a worldwide scale [85, 86]. We may anticipate new developments as this field of study develops that take advantage of the remarkable qualities of nano cellulose, providing a greener and more sustainable energy future [87, 88].

5. Future directions and challenges

- The search for nature-inspired nanocellulose materials, improvements in manufacturing processes, and a wide range of uses is an ongoing adventure that offers both thrilling possibilities and difficult obstacles. The trajectory of research and innovation in this discipline is likely to be shaped by several future objectives and obstacles as the field matures.

- Nanocellulose qualities, like mechanical strength, thermal stability, and electrical conductivity, will likely be the subject of future research that aims to be more precisely tailored to satisfy the needs of certain applications.

- The development of precise modification techniques, including surface functionalization and composite formulations, will enable fine-tuning of these properties.

5.1 Emerging trends in nano cellulose research

Several new trends that are emerging are influencing the direction that nanocellulose research will take as it continues to develop. Researchers are concentrating on creating unique nano cellulose variants with specialized features for certain applications, such as cellulose nanocrystals and cellulose nanofibrils. Techniques for surface functionalization and modification are becoming more popular to increase compatibility with various matrices and boost performance. Additionally, multidisciplinary partnerships between engineers, biotechnologists, and material scientists are increasingly widespread, resulting in ground-breaking discoveries and multifunctional materials. By combining nanocellulose with other cutting-edge materials like graphene and nanoparticles, new opportunities for synergistic effects and improved functionality are becoming possible. To develop materials with superior mechanical characteristics and self-healing capacities, efforts are also focused on comprehending and imitating nature's hierarchical structures. Another intriguing area that shows promise for the future is investigating the possibilities of nanocellulose in the realm of nanomedicine, including targeted medication delivery and tissue engineering.

5.2 Addressing scalability and cost-effectiveness

Although nano cellulose has great potential, scaling and cost-effectiveness issues make it difficult to use on a large scale. Large-scale manufacturing of nano cellulose is economically difficult since it frequently includes energy-intensive processes and specialized equipment. To increase sustainability and lower costs, researchers are actively striving to improve production techniques, create more energy-efficient processes, and investigate the use of waste streams as feedstock. Production expansion is still a top aim as long as constant quality and minimal environmental impact are maintained. To create affordable and scalable processes for producing nanocellulose, collaboration between government, business, and academia is essential. Additionally, investigating bioengineering techniques to increase cellulose yield in plants or microorganisms may provide creative ideas to get around scalability obstacles.

5.3 Environmental considerations and life cycle analysis

Addressing the environmental impact of nano cellulose production and use is of utmost importance as the demand for sustainable materials rises. To examine the entire environmental impact of nano cellulose materials and pinpoint any potential problem areas in their manufacturing, use, and disposal, thorough life cycle assessments (LCAs) must be carried out. To lessen the usage of harsh chemicals and energy-intensive stages, researchers are actively investigating greener and more environmentally friendly extraction technologies, such as enzyme-assisted processes. For nanocellulose-based goods, recycling and circular economy strategies are being researched to reduce waste generation and advance resource efficiency. For responsible development, it is also essential to comprehend the potential environmental dangers connected to nanocellulose materials, such as their behavior in aquatic systems and potential ecotoxicity. To ensure the safe and sustainable usage of nanocellulose across a range of applications, regulatory frameworks and guidelines may need to be established.

5.4 Holistic approaches and collaborative initiatives

A comprehensive strategy that incorporates scientific research, engineering innovation, commercial viability, and environmental stewardship is needed to address the issues and realize the full potential of nano cellulose materials. Knowledge exchange, technology transfer, and the creation of standardized protocols can be facilitated by collaborative projects involving academia, business, governmental organizations, and non-governmental organizations. Openly exchanging data, processes, and best practices can hasten development and reduce effort duplication. International partnerships can promote a global perspective on research into nanocellulose by allowing the sharing of various perspectives and experiences from various places. The multidisciplinary nature of nano cellulose research necessitates global collaboration and knowledge exchange among scientists, engineers, and stakeholders from different regions. Open-access platforms, conferences, and research networks will play a crucial role in facilitating collaboration and accelerating advancements. In conclusion, the future of nano cellulose materials holds great promise, with emerging trends pushing the boundaries of possibilities. Addressing challenges related to scalability, cost-effectiveness, and environmental sustainability is essential for unlocking the full potential of these nature-inspired materials and realizing their transformative impact across various industries. By adopting a multidisciplinary and collaborative approach, researchers and stakeholders can drive innovation and shape a more sustainable future.

6. Conclusions

From the nature perspective, the current study summarizes the available resources for extraction of nano cellulose, and their extraction methods with the potential applications. Based on the current research trend, the scientific community identified the nano cellulose and existed from forest biomass, agro waste, and ocean biomass, for many applications in Biomedical, aerospace, energy, pharmaceutical and other versatile applications in emerging areas. The economical, thermal, mechanical, morphological, and chemical properties of the nano cellulose broadened the applications where it fit. Since there is huge demand of nano cellulose materials for sustainable applications is indeed requirement for large production. Also, nano cellulose materials are increasing the economic aspects of productivity, and it is playing a very important role in environmental safety and protection.

The quest for efficient and eco-friendly nano cellulose preparation methods has been guided by nature's ingenious processes. Biomimetic extraction techniques, mechanical disintegration, and self-assembly have all contributed to unlocking the potential of nano cellulose. Understanding the structural, mechanical, and thermal characteristics of nanocellulose materials has benefited from thorough characterization. Applications of nature-inspired nano cellulose materials include biomedical devices, textiles, water purification, and energy solutions in addition to sustainable packaging. A more sustainable and technologically advanced future is now possible thanks to the extraordinary qualities of nano cellulose, such as its mechanical strength, thermal stability, and biocompatibility.

Nanocellulose materials with natural inspiration have been instrumental in the development of sustainable energy systems. Their use in flexible electronics, energy storage systems, batteries, and capacitors has improved performance, increased the efficiency of energy storage, and made it possible to develop more ecologically friendly solutions for energy storage and utilization. In conclusion, the development of nanocellulose materials that are inspired by nature, improvements in the methods used to prepare them, and their wide range of applications highlight the incredible promise of biomimetic methods in materials research. This journey of discovery stands as a testament to the ingenuity of researchers and their commitment to harnessing nature's wisdom to address contemporary challenges. As this field continues to evolve, we anticipate even greater strides in sustainable innovation, offering a brighter and more sustainable future for generations to come.

Conflict of interest

The authors declare no conflict of interest.

Author details

Hanumanthu Jeevan Rao[1*], Sanjay Singh[1], Perumalla Janaki Ramulu[2], Narender Singh[1], Thiago F. Santos[3], Caroliny M. Santos[3], Nandini Robin Nadar[4] and Gara Dheeraj Kumar[4]

1 Amity Institute of Aerospace Engineering, Amity University Uttar Pradesh, Noida, Uttar Pradesh, India

2 Department of Mechanical Engineering, School of Mechanical, Chemical and Materials Engineering, ASTU, Adama, Ethiopia

3 Postgraduate Program in Chemical Engineering, Technology Center, Federal University of Rio Grande do Norte, Brazil

4 Department of Aerospace, Jain (Deemed-to-be University), Bengaluru, Karnataka, India

*Address all correspondence to: jeevan.rao14@gmail.com

IntechOpen

© 2024 The Author(s). Licensee IntechOpen. This chapter is distributed under the terms of the Creative Commons Attribution License (http://creativecommons.org/licenses/by/3.0), which permits unrestricted use, distribution, and reproduction in any medium, provided the original work is properly cited.

References

[1] Nagarajan KJ, Ramanujam NR, Sanjay MR, Siengchin S, Surya Rajan B, Sathick Basha K, et al. A comprehensive review on cellulose nanocrystals and cellulose nanofibers: Pretreatment, preparation, and characterization. Polymer Composites. 2021;**42**:1588-1630. DOI: 10.1002/pc.25929

[2] Syafri E, Kasim A, Abral H, Sudirman GT, Sulungbudi MR, SanjayNHS.Synthesisandcharacterization of cellulose nanofibers (CNF) ramie reinforced cassava starch hybrid composites. International Journal of Biological Macromolecules. 2018;**120**:578-586. DOI: 10.1016/j.ijbiomac.2018.08.134

[3] Trache D, Tarchoun AF, Derradji M, Hamidon TS, Masruchin N, Brosse N, et al. Nanocellulose: From fundamentals to advanced applications. Frontiers in Chemistry. 2020;**8**. DOI: 10.3389/fchem.2020.00392. Available from: https://www.frontiersin.org/articles/10.3389/fchem.2020.00392/full

[4] Shanmugam K, Doosthosseini H, Varanasi S, Garnier G, Batchelor W. Nanocellulose films as air and water vapour barriers: A recyclable and biodegradable alternative to polyolefin packaging. Sustainable Materials and Technologies. 2019;**22**:e00115. DOI: 10.1016/j.susmat.2019.e00115

[5] Zubair M, Arshad M, Ullah A. Nanocellulose: A sustainable and renewable material for water and wastewater treatment. In: Natural Polymers-Based Green Adsorbents for Water Treatment. Elsevier; 2021. pp. 93-109. DOI: 10.1016/B978-0-12-820541-9.00009-0. Available from: https://www.frontiersin.org/articles/10.3389/fchem.2020.601256/full

[6] Gan PG, Sam ST, bin Abdullah MF, Omar MF. Thermal properties of nanocellulose-reinforced composites: A review. Journal of Applied Polymer Science. 2020;**137**:48544. DOI: 10.1002/app.48544

[7] Thakur MK, Thakur VK, Prasanth R. Nanocellulose-based polymer nanocomposites: An introduction. In: Nanocellulose Polymer Nanocomposites. Hoboken, NJ, USA: John Wiley & Sons, Inc.; 2014. pp. 1-15. DOI: 10.1002/9781118872246.ch1

[8] Thomas P, Duolikun T, Rumjit NP, Moosavi S, Lai CW, Bin Johan MR, et al. Comprehensive review on nanocellulose: Recent developments, challenges and future prospects. Journal of the Mechanical Behavior of Biomedical Materials. 2020;**110**:103884. DOI: 10.1016/j.jmbbm.2020.103884

[9] Dhali K, Ghasemlou M, Daver F, Cass P, Adhikari B. A review of nanocellulose as a new material towards environmental sustainability. Science of the Total Environment. 2021;**775**:145871. DOI: 10.1016/j.scitotenv.2021.145871

[10] Barhoum A, Rastogi VK, Mahur BK, Rastogi A, Abdel-Haleem FM, Samyn P. Nanocelluloses as new generation materials: Natural resources, structure-related properties, engineering nanostructures, and technical challenges. Materials Today Chemistry. 2022;**26**:101247. DOI: 10.1016/j.mtchem.2022.101247

[11] Park N-M, Choi S, Oh JE, Hwang DY. Facile extraction of cellulose nanocrystals. Carbohydrate Polymers. 2019;**223**:115114. DOI: 10.1016/j.carbpol.2019.115114

[12] Pradhan D, Jaiswal AK, Jaiswal S. Emerging technologies for

the production of nanocellulose from lignocellulosic biomass. Carbohydrate Polymers. 2022;**285**:119258. DOI: 10.1016/j.carbpol.2022.119258

[13] Thomas B, Raj MC, Joy J, Moores A, Drisko GL, Sanchez C. Nanocellulose, a versatile green platform: From biosources to materials and their applications. Chemical Reviews. 2018;**118**:11575-11625. DOI: 10.1021/acs.chemrev.7b00627

[14] He Q, Yang Y, Liu Z, Shao D, Jiang D, Xing L, et al. Preparation and characterization of cellulose nanocrystals from spent edible fungus substrate. Journal of the Science of Food and Agriculture. 2022;**102**:2761-2772. DOI: 10.1002/jsfa.11617

[15] Mateo S, Peinado S, Morillas-Gutiérrez F, La Rubia MD, Moya AJ. Nanocellulose from agricultural wastes: Products and applications—A review. PRO. 2021;**9**:1594. DOI: 10.3390/pr9091594

[16] de Souza AG, Kano FS, Bonvent JJ, dos S. Rosa D. Cellulose nanostructures obtained from waste paper industry: A comparison of acid and mechanical isolation methods. Materials Research. 2017;**20**:209-214. DOI: 10.1590/1980-5373-mr-2016-0863

[17] Yu S, Sun J, Shi Y, Wang Q, Wu J, Liu J. Nanocellulose from various biomass wastes: Its preparation and potential usages towards the high value-added products. Environmental Science & Ecotechnology. 2021;**5**:100077. DOI: 10.1016/j.ese.2020.100077

[18] Tayeb A, Amini E, Ghasemi S, Tajvidi M. Cellulose nanomaterials—Binding properties and applications: A review. Molecules. 2018;**23**:2684. DOI: 10.3390/molecules23102684

[19] Randhawa A, Dutta SD, Ganguly K, Patil TV, Patel DK, Lim K-T. A review of properties of nanocellulose, its synthesis, and potential in biomedical applications. Applied Sciences. 2022;**12**:7090. DOI: 10.3390/app12147090

[20] Sharma A, Thakur M, Bhattacharya M, Mandal T, Goswami S. Commercial application of cellulose nano-composites – A review. Biotechnology Reports. 2019;**21**:e00316. DOI: 10.1016/j.btre.2019.e00316

[21] Yu K, Cao H, Qian K, Sha X, Chen Y, Haijian C. Shear-thickening behavior of modified silica nanoparticles in polyethylene glycol. Journal of Nanoparticle Research. 2012;**14**:1-9. DOI: 10.1007/s11051-012-0747-2

[22] Gupta GK, Shukla P. Lignocellulosic biomass for the synthesis of nanocellulose and its eco-friendly advanced applications. Frontiers in Chemistry. 2020;**8**. DOI: 10.3389/fchem.2020.601256

[23] Owonubi SJ, Agwuncha SC, Malima NM, Shombe GB, Makhatha EM, Revaprasadu N. Non-woody biomass as sources of nanocellulose particles: A review of extraction procedures. Frontiers in Energy Research. 2021;**9**. DOI: 10.3389/fenrg.2021.608825

[24] Nargotra P, Sharma V, Tsai M-L, Hsieh S-L, Dong C-D, Wang H-MD, et al. Recent advancements in the valorization of Agro-industrial food waste for the production of nanocellulose. Applied Sciences. 2023;**13**:6159. DOI: 10.3390/app13106159

[25] Lahiri A, Daniel S, Kanthapazham R, Vanaraj R, Thambidurai A, Peter LS. A critical review on food waste management for the production of materials and biofuel. Journal of Hazardous Materials Advances. 2023;**10**:100266. DOI: 10.1016/j.hazadv.2023.100266

[26] Aljabali AAA, Obeid MA, Rezigue MM, Alqudah A, Charbe NB, Chellappan DK, et al. Nanocelluloses as a novel vehicle for controlled drug delivery. In: Handbook of Nanocellulose. Cham: Springer International Publishing; 2022. pp. 507-542. DOI: 10.1007/978-3-030-89621-8_36

[27] Thakur V, Guleria A, Kumar S, Sharma S, Singh K. Recent advances in nanocellulose processing, functionalization and applications: A review. Materials Advances. 2021;**2**:1872-1895. DOI: 10.1039/D1MA00049G

[28] Eichhorn SJ, Etale A, Wang J, Berglund LA, Li Y, Cai Y, et al. Current international research into cellulose as a functional nanomaterial for advanced applications. Journal of Materials Science. 2022;**57**:5697-5767. DOI: 10.1007/s10853-022-06903-8

[29] De Melo KM, Dos Santos TF, Santos CMDS, Da Fonseca RT, De Lucena ND, De Medeiros JI, et al. Study of the reuse potential of the sisal fibers powder as a particulate material in polymer composites. Journal of Materials Research and Technology. 2019;**8**:4019-4025. DOI: 10.1016/j.jmrt.2019.07.010

[30] Santos C, Santos T, Aquino M, Alves S. Extraction and characterization of cellulosic fibers from the stem of papaya tree (Carica papaya L.). In: Ramnath R, Rangappa SM, Siengchin S, Fiore V, editors. Cellulose Fibre Reinforced Composites. Interface Engineering, Processing and Performance. 1st ed. Elsevier; 2022. p. 434. DOI: 10.1016/B978-0-323-90125-3.00010-0. Available from: https://www.frontiersin.org/articles/10.3389/fenrg.2021.608825/full

[31] Rangappa SM, Siengchin S, Parameswaranpillai J, Jawaid M, Ozbakkaloglu T. Lignocellulosic fiber reinforced composites: Progress, performance, properties, applications, and future perspectives. Polymer Composites. 2022;**43**:645-691. Available from: https://www.frontiersin.org/articles/10.3389/fenrg.2021.608825/full. DOI: 10.1002/PC.26413

[32] Vinod A, Sanjay MR, Suchart S, Jyotishkumar P. Renewable and sustainable biobased materials: An assessment on biofibers, biofilms, biopolymers and biocomposites. Journal of Cleaner Production. 2020;**258**:120978. DOI: 10.1016/j.jclepro.2020.120978

[33] Santos C, Santos T, Manicoba E, Aquino M. Biofibers of papaya tree bast: A statistical study of the mechanical properties for use potential in polymeric composites. In: Advances in Bio-Based Fibres: Moving Towards a Green Society. 2022. pp. 687-705. Available from: https://link.springer.com/article/10.1007/s13399-023-04513-1. DOI: 10.1016/B978-0-12-824543-9.00031-1

[34] Santos CM, Santos TF, Rangappa SM, Macedo M, Aquino M, Alves SM, et al. Physical & mechanical and chemical properties on papaya tree bast fibers from different portions. Biomass Conversion and Biorefinery. 2023. DOI: 10.1007/s13399-023-04513-1. Available from: https://sdgresources.relx.com/bookchapters/natural-polymers-based-greenadsorbents-water-treatment-chapter-4-nanocellulose

[35] Santos C, Santos T, Moreira K, Aquino M, Zillio RFL. Statistical study of the influence of Fiber content, Fiber length and critical length in the mechanical behavior of polymeric composites reinforced with Carica Papaya Fibers (CPFs). Applied Science and Engineering Progress. 2021;**14**:719-726. DOI: 10.14416/J.ASEP.2021.07.002

[36] Karimah A, Ridho MR, Munawar SS, Adi DS, Ismadi R, Damayanti B, et al. A

review on natural fibers for development of eco-friendly bio-composite: Characteristics, and utilizations. Journal of Materials Research and Technology. 2021;**13**:2442-2458. DOI: 10.1016/j.jmrt.2021.06.014

[37] Gurunathan T, Mohanty S, Nayak SK. A review of the recent developments in biocomposites based on natural fibres and their application perspectives. Composites. Part A, Applied Science and Manufacturing. 2015;**77**:1-25. DOI: 10.1016/j.compositesa.2015.06.007

[38] Wang J, Han X, Zhang C, Liu K, Duan G. Source of nanocellulose and its application in nanocomposite packaging material: A review. Nanomaterials. 2022;**12**:3158. DOI: 10.3390/nano12183158

[39] Jasmani L, Adnan S. Preparation and characterization of nanocrystalline cellulose from Acacia mangium and its reinforcement potential. Carbohydrate Polymers. 2017;**161**:166-171. DOI: 10.1016/j.carbpol.2016.12.061

[40] García A, Labidi J, Belgacem MN, Bras J. The nanocellulose biorefinery: Woody versus herbaceous agricultural wastes for NCC production. Cellulose. 2017;**24**:693-704. DOI: 10.1007/s10570-016-1144-2

[41] Rasheed M, Jawaid M, Parveez B, Zuriyati A, Khan A. Morphological, Chemical and Thermal Analysis of Cellulose Nanocrystals Extracted from Bamboo Fibre. Elsevier B.V; 2020. DOI: 10.1016/j.ijbiomac.2020.05.170 Available from: https://bioresourcesbioprocessing.springeropen.com/articles/10.1186/s40643-019-0279-z

[42] Phanthong P, Karnjanakom S, Reubroycharoen P, Hao X, Abudula A, Guan G. A facile one-step way for extraction of nanocellulose with high yield by ball milling with ionic liquid. Cellulose. 2017;**24**:2083-2093. DOI: 10.1007/s10570-017-1238-5

[43] Kharismi RRAY, Sutriyo H. Suryadi, preparation and characterization of microcrystalline cellulose produced from betung bamboo (dendrocalamus asper) through acid hydrolysis. Journal of Young Pharmacists. 2018;**10**:s79-s83. DOI: 10.5530/jyp.2018.2s.15

[44] El Achaby M, El Miri N, Aboulkas A, Zahouily M, Bilal E, Barakat A, et al. Processing and properties of eco-friendly bio-nanocomposite films filled with cellulose nanocrystals from sugarcane bagasse. International Journal of Biological Macromolecules. 2017;**96**:340-352. DOI: 10.1016/j.ijbiomac.2016.12.040

[45] Liu Y, Wang H, Yu G, Yu Q, Li B, Mu X. A novel approach for the preparation of nanocrystalline cellulose by using phosphotungstic acid. Carbohydrate Polymers. 2014;**110**:415-422. DOI: 10.1016/j.carbpol.2014.04.040

[46] Zhang Y, Xu Y, Yue X, Dai L, Ni Y. Isolation and characterization of microcrystalline cellulose from bamboo pulp through extremely low acid hydrolysis. Journal of Wood Chemistry and Technology. 2019;**39**:242-254. DOI: 10.1080/02773813.2019.1566365

[47] Bufalino L, Tonoli GHD, Costa TG, Protásio TP, Neto ARS, Marconcini JM, et al. Nanocellulose Films from Amazon Forest wood wastes: Structural and thermal properties. In: Non-Conventional Materials and Technologies for Sustainable Development. Trans Tech Publications Ltd; 2016. pp. 110-117. DOI: 10.4028/www.scientific.net/KEM.668.110. Available from: https://www.sciencedirect.com/science/article/pii/S0141813020328464?via%3Dihub

[48] Song K, Zhu X, Zhu W, Li X. Preparation and characterization of

cellulose nanocrystal extracted from Calotropis procera biomass. Bioresources and Bioprocessing. 2019;**6**. DOI: 10.1186/s40643-019-0279-z

[49] Abu-Thabit NY, Judeh AA, Hakeem AS, Ul-Hamid A, Umar Y, Ahmad A. Isolation and characterization of microcrystalline cellulose from date seeds (Phoenix dactylifera L.). International Journal of Biological Macromolecules. 2020;**155**:730-739. DOI: 10.1016/j.ijbiomac.2020.03.255

[50] Vallejos ME, Felissia FE, Area MC, Ehman NV, Tarrés Q, Mutjé P. Nanofibrillated cellulose (CNF) from eucalyptus sawdust as a dry strength agent of unrefined eucalyptus handsheets. Carbohydrate Polymers. 2016;**139**:99-105. DOI: 10.1016/j.carbpol.2015.12.004

[51] Baruah J, Deka RC, Kalita E. Greener production of microcrystalline cellulose (MCC) from Saccharum spontaneum (Kans grass): Statistical optimization. International Journal of Biological Macromolecules. 2020;**154**:672-682. DOI: 10.1016/j.ijbiomac.2020.03.158

[52] Hasan M, Lai TK, Gopakumar DA, Jawaid M, Owolabi FAT, Mistar EM, et al. Micro crystalline bamboo cellulose based seaweed biodegradable composite films for sustainable packaging material. Journal of Polymers and the Environment. 2019;**27**:1602-1612. DOI: 10.1007/s10924-019-01457-4

[53] Moriana R, Vilaplana F, Ek M. Cellulose Nanocrystals from Forest residues as reinforcing agents for composites: A study from macro- to Nano-dimensions. Carbohydrate Polymers. 2016;**139**:139-149. DOI: 10.1016/j.carbpol.2015.12.020

[54] Couret L, Irle M, Belloncle C, Cathala B. Extraction and characterization of cellulose nanocrystals from post-consumer wood fiberboard waste. Cellulose. 2017;**24**:2125-2137. DOI: 10.1007/s10570-017-1252-7

[55] Zianor Azrina ZA, Beg MDH, Rosli MY, Ramli R, Junadi N, Alam AKMM. Spherical nanocrystalline cellulose (NCC) from oil palm empty fruit bunch pulp via ultrasound assisted hydrolysis. Carbohydrate Polymers. 2017;**162**:115-120. DOI: 10.1016/j.carbpol.2017.01.035

[56] Rohaizu R, Wanrosli WD. Sono-assisted TEMPO oxidation of oil palm lignocellulosic biomass for isolation of nanocrystalline cellulose. Ultrasonics Sonochemistry. 2017;**34**:631-639. DOI: 10.1016/j.ultsonch.2016.06.040

[57] Xiao S, Gao R, Lu Y, Li J, Sun Q. Fabrication and characterization of nanofibrillated cellulose and its aerogels from natural pine needles. Carbohydrate Polymers. 2015;**119**:202-209. DOI: 10.1016/j.carbpol.2014.11.041

[58] García-García D, Balart R, Lopez-Martinez J, Ek M, Moriana R. Optimizing the yield and physico-chemical properties of pine cone cellulose nanocrystals by different hydrolysis time. Cellulose. 2018;**25**:2925-2938. DOI: 10.1007/s10570-018-1760-0

[59] Rambabu N, Panthapulakkal S, Sain M, Dalai AK. Production of nanocellulose fibers from pinecone biomass: Evaluation and optimization of chemical and mechanical treatment conditions on mechanical properties of nanocellulose films. Industrial Crops and Products. 2016;**83**:746-754. DOI: 10.1016/j.indcrop.2015.11.083

[60] Tuerxun D, Pulingam T, Nordin NI, Chen YW, Bin Kamaldin J, Julkapli NBM, et al. Synthesis, characterization and cytotoxicity studies of nanocrystalline cellulose from the production waste of rubber-wood and kenaf-bast

fibers. European Polymer Journal. 2019;**116**:352-360. DOI: 10.1016/j.eurpolymj.2019.04.021

[61] Abdul Khalil HPS, Lai TK, Tye YY, Paridah MT, Fazita MRN, Azniwati AA, et al. Preparation and characterization of microcrystalline cellulose from sacred Bali bamboo as reinforcing filler in seaweed-based composite film. Fibers and Polymers. 2018;**19**:423-434. DOI: 10.1007/s12221-018-7672-7

[62] Blasi A, Verardi A, Lopresto CG, Siciliano S, Sangiorgio P. Lignocellulosic agricultural waste valorization to obtain valuable products: An overview. Recycling. 2023;**8**:61. DOI: 10.3390/recycling8040061

[63] Wu F, Misra M, Mohanty AK. Challenges and new opportunities on barrier performance of biodegradable polymers for sustainable packaging. Progress in Polymer Science. 2021;**117**:101395. DOI: 10.1016/j.progpolymsci.2021.101395

[64] Fang Z, Hou G, Chen C, Hu L. Nanocellulose-based films and their emerging applications. Current Opinion in Solid State & Materials Science. 2019;**23**:100764. DOI: 10.1016/j.cossms.2019.07.003

[65] Norizan MN, Shazleen SS, Alias AH, Sabaruddin FA, Asyraf MRM, Zainudin ES, et al. Nanocellulose-based nanocomposites for sustainable applications: A review. Nanomaterials. 2022;**12**:3483. DOI: 10.3390/nano12193483

[66] Fujisawa S. Material design of nanocellulose/polymer composites via Pickering emulsion templating. Polymer Journal. 2021;**53**:103-109. DOI: 10.1038/s41428-020-00408-4

[67] Ghilan A, Nicu R, Ciolacu DE, Ciolacu F. Insight into the latest medical applications of nanocellulose. Materials (Basel). 2023;**16**:4447. DOI: 10.3390/ma16124447

[68] Chakrabarty A, Teramoto Y. Recent advances in nanocellulose composites with polymers: A guide for choosing partners and how to incorporate them. Polymers (Basel). 2018;**10**:517. DOI: 10.3390/polym10050517

[69] Omran AAB, Mohammed AABA, Sapuan SM, Ilyas RA, Asyraf MRM, Rahimian Koloor SS, et al. Micro- and nanocellulose in polymer composite materials: A review. Polymers (Basel). 2021;**13**:231. DOI: 10.3390/polym13020231

[70] George B, Lal N, Suchithra TV. Nanocellulose as polymer composite reinforcement material. In: Plant Nanobionics: Volume 2, Approaches in Nanoparticles, Biosynthesis, and Toxicity. 2019. pp. 409-427. DOI: 10.1007/978-3-030-16379-2_14. Available from: https://link.springer.com/book/10.1007/978-3-030-16379-2

[71] Lee K-Y, Aitomäki Y, Berglund LA, Oksman K, Bismarck A. On the use of nanocellulose as reinforcement in polymer matrix composites. Composites Science and Technology. 2014;**105**:15-27. DOI: 10.1016/j.compscitech.2014.08.032

[72] Khalid MY, Al Rashid A, Arif ZU, Ahmed W, Arshad H. Recent advances in nanocellulose-based different biomaterials: Types, properties, and emerging applications. Journal of Materials Research and Technology. 2021;**14**:2601-2623. DOI: 10.1016/j.jmrt.2021.07.128

[73] Padzil FNM, Lee SH, Ainun ZMA, Lee CH, Abdullah LC. Potential of oil palm empty fruit bunch resources in nanocellulose hydrogel production for versatile applications: A review. Materials

(Basel). 2020;**13**:1245. DOI: 10.3390/ma13051245

[74] Moreira K, Santos T, Santos C, Fonseca R, Melo M, Aquino M. Analysis of the physical and mechanical properties of a biobased composite with sisal powder. In: Biobased Composites. Wiley; 2021. pp. 143-151. DOI: 10.1002/9781119641803.ch10. Available from: https://www.scientific.net/KEM.668.110

[75] Melo K, Santos T, Santos C, Fonseca R, Dantas N, Aquino M. Experimental analysis of styrene, particle size, and Fiber content in the mechanical properties of sisal Fiber powder composites. In: Hybrid Fiber Composites. Wiley; 2020. pp. 351-367. DOI: 10.1002/9783527824571.ch16. Available from: https://onlinelibrary.wiley.com/doi/10.1002/9781119641803ch10

[76] Santos T, Santos C, Fonseca R, Melo K, Aquino M. Natural fibres based phenolic composites. In: Phenolic Polymers Based Composite Materials. Singapore: Springer Singapore; 2021. pp. 65-75. DOI: 10.1007/978-981-15-8932-4_4

[77] Gumrah Dumanli A. Nanocellulose and its composites for biomedical applications. Current Medicinal Chemistry. 2017;**24**:512-528. DOI: 10.2174/092986732 3666161014124008

[78] Nicu R, Ciolacu F, Ciolacu DE. Advanced functional materials based on nanocellulose for pharmaceutical/medical applications. Pharmaceutics. 2021;**13**:1125. DOI: 10.3390/pharmaceutics13081125

[79] Mishra PK, Pavelek O, Rasticova M, Mishra H, Ekielski A. Nanocellulose-based biomedical scaffolds in future bioeconomy: A techno-legal assessment of the state-of-the-art. Frontiers in Bioengineering and Biotechnology. 2022;**9**. DOI: 10.3389/fbioe.2021.789603

[80] Nehra P, Chauhan RP. Eco-friendly nanocellulose and its biomedical applications: Current status and future prospect. Journal of Biomaterials Science. Polymer Edition. 2021;**32**:112-149. DOI: 10.1080/09205063.2020.1817706

[81] Voisin H, Bergström L, Liu P, Mathew A. Nanocellulose-based materials for water purification. Nanomaterials. 2017;7:57. DOI: 10.3390/nano7030057

[82] Salama A, Abouzeid R, Leong WS, Jeevanandam J, Samyn P, Dufresne A, et al. Nanocellulose-based materials for water treatment: Adsorption, photocatalytic degradation, disinfection, antifouling, and nanofiltration. Nanomaterials. 2021;**11**:3008. DOI: 10.3390/nano11113008

[83] Xiao J, Li H, Zhang H, He S, Zhang Q, Liu K, et al. Nanocellulose and its derived composite electrodes toward supercapacitors: Fabrication, properties, and challenges. Journal of Bioresources and Bioproducts. 2022;7:245-269. DOI: 10.1016/j.jobab.2022.05.003

[84] Li R, Tian D, Chen L, Zhuang B, Feng H, Li Q, et al. The application of cellulose nanofibrils in energy systems. Batteries. 2023;**9**:399. DOI: 10.3390/batteries9080399

[85] Chen W, Yu H, Lee S-Y, Wei T, Li J, Fan Z. Nanocellulose: A promising nanomaterial for advanced electrochemical energy storage. Chemical Society Reviews. 2018;**47**:2837-2872. DOI: 10.1039/C7CS00790F

[86] Ji SM, Kumar A. Cellulose-derived nanostructures as sustainable biomass for supercapacitors: A review. Polymers

(Basel). 2022;**14**:169. DOI: 10.3390/polym14010169

[87] Maity CK, De S, Verma K, Moniruzzaman M, Sahoo S. Nanocellulose: A versatile nanostructure for energy storage applications. Industrial Crops and Products. 2023;**204**:117218. DOI: 10.1016/j.indcrop.2023.117218

[88] Chen C, Hu L. Nanocellulose toward advanced energy storage devices: Structure and electrochemistry. Accounts of Chemical Research. 2018;**51**:3154-3165. DOI: 10.1021/acs.accounts.8b00391

Chapter 4

Fascinating Properties and Applications of Nanocellulose in the Food Industry

Elham Asghari-Varzaneh and Hajar Shekarchizadeh

Abstract

Nanocellulose, a material derived from cellulose fibers, has gained significant attention in various industries due to its unique properties and potential applications. From an economic perspective, using nanocellulose in industry offers several advantages, such as cost-effectiveness, enhanced product performance, environmental benefits, and diversified applications. Also, it is utilized in the food industry because of its distinct properties, including high surface area, rheological behavior, water absorption ability, crystallinity, and no cytotoxicity. A significant application of nanocellulose is its potential to replace fats, carbohydrates, and proteins and serve as stabilizing agents in high-calorie foods. Moreover, nanocellulose has demonstrated exceptional efficacy as a delivery system, making it an ideal choice for preserving nutrients and active ingredients in food products. A primary objective in the packaging industry is to maintain food quality, extend its shelf life, and minimize waste. Since nanocellulose is both renewable and natural and offers oxygen and water vapor barrier properties, it emerges as a suitable candidate for the packaging industry. However, despite its promising features and applications, there are uncertainties around its non-toxicity and the potential impact on human health, issues that are currently being examined by scientists in the food industry.

Keywords: nanocellulose, food stabilizer, dietary fiber, encapsulating agent, packaging

1. Introduction

Cellulose $(C_6H_{10}O_5)_n$, one of the most abundant natural polymers in the world, was first discovered by Anselme Payene. Cellulose is the main component of the cell walls of most plants. The main sources of cellulose are wood pulp, cotton stalks, straw, bacteria, sea animals, and agricultural waste (fruits and vegetables) [1]. Renewability, non-toxicity, environmentally-friendly, wide availability, excellent mechanical properties, low weight, reinforcing capabilities, and biodegradability, with no adverse effects on health and environment, are among the important and practical features of cellulose used in various industries such as food, paper, bio-material, and pharmaceutical industries [2, 3]. Cellulose is a homopolysaccharide compound formed by joining large numbers of β-D-glucose monomers with β (1,4) glycosidic bonds [4]. Cellulose forms intra- and intermolecular hydrogen bonds due to hydroxyl groups in its structure. In general, these hydroxyl groups and their ability to create hydrogen bonds affect the biodegradability, chirality, and hydrophilic properties of cellulose and also create important functions, such as microfibrillated structure, hierarchical organization (crystalline and amorphous fractions), and cohesive nature [5, 6].

Cellulose derivatives, including cellulose acetate, cellulose sulfate, cellulose nitrate, carboxymethyl cellulose, ethyl cellulose, methylcellulose, and nanocellulose, exhibit distinct properties (sizes and shapes) and characteristics based on the production method and materials employed [1].

Nanocellulose consists of cellulose particles with dimensions less than 100 nm. The nanocellulose fibers can be obtained through chemical modification, such as acid hydrolysis, or physical modification, such as ultrasonication [7]. Nanocellulose possesses several practical and significant properties that make it an ideal choice for preparing various polymer composites, including low density, non-abrasiveness, combustibility, excellent potential for chemical modification, crystalline structure, increased specific surface area, advantageous rheological properties, outstanding resistance to water vapor transmission and oxygen permeation, effective barrier against aqueous liquids and oil grease, non-toxicity, and inexpensiveness compared to other synthetic polymers [8, 9].

Nanocellulose, an environmentally friendly and biodegradable biomaterial, has emerged as a promising alternative to synthetic materials. In recent years, cellulose nanocomposites have found applications in various industries, including food, medical, biomedical, pharmaceutical, civil construction, automotive, electronics, packaging, construction, and wastewater treatment [8, 10].

For example, nanocellulose-based biomaterials have properties similar to natural tissue, making them proper for cell attachment and growth. Nanocrystalline suspensions can be used as a cell culture environment, and cellulose nanofibers have been studied to grow membranes containing healing and bacteriostatic agents for treating skin burns [11]. Also, in the field of application in the construction industry, it is possible to mention the efforts of researchers to produce products compatible with the environment. For example, in 2019, a study was conducted to investigate the mechanical performance of integrating cellulose fibers into cement systems. The findings showed that the material had improved properties, including compressive strength, modulus of elasticity, and toughness, indicating its potential as a promising material with ductile and more resistant behavior [12].

2. Economic aspects of using nanocellulose in industry

Nanocellulose can be produced from different raw materials, which depend on the availability of the raw material, its cost, and environmental conditions, and these factors can directly affect the cost of nanocellulose production. This is observed, even though the properties of the produced nanocellulose and its application in the industry are influenced by the raw material and production method. As research and innovation in nanocellulose-based technologies continue, new sources and production methods are expected to be recognized and developed, further expanding its potential applications over the years [13]. In general, the economic advantages of the use of nanocellulose in industry can be mentioned as follows:

1. Cost-effectiveness: nanocellulose can be produced from renewable sources such as wood pulp, agricultural residues, or waste paper. This makes it a cost-effective and supportable alternative to customary materials. Additionally, the production

processes for nanocellulose are becoming more efficient, further reducing production costs; 2. Lightweight and strong: nanocellulose is lightweight yet incredibly strong, making it an attractive material for industries such as automotive, aerospace, and construction. Its high strength-to-weight ratio can lead to cost savings in manufacturing and transportation; 3. Enhanced product performance: the interpolation of nanocellulose into various products can improve their mechanical strength, barrier properties, and persistence. This can result in longer-lasting products with reduced maintenance and replacement costs; 4. Environmental benefits: nanocellulose is biodegradable and non-toxic, making it an environmentally friendly alternative to synthetic materials. As sustainability becomes increasingly important to consumers and regulatory bodies, the use of nanocellulose can provide a competitive advantage for companies looking to reduce their environmental impact; 5. Variegated applications: nanocellulose has a wide range of potential applications across industries such as packaging, textiles, electronics, healthcare, and energy storage. This versatility opens up new market opportunities for companies utilizing nanocellulose in their products.

Apart from the economic benefits of using nanocellulose in the industry, there are several disadvantages and barriers facing nanocellulose and its wide application in various industries, which can be mentioned as 1. Initial investment: implementing and developing efficient and valuable methods of nanocellulose production can be expensive, which is the biggest challenge for manufacturing companies; 2. Competition: nanocellulose faces competition from other materials, and companies may need to invest in marketing and education efforts to promote its advantages and gain market share; 3. Uncertain demand: due to the novelty of using nanocellulose in the industry, this material may face uncertainty in the market and price; 4. Regulatory challenges: regulatory requirements and standards for nanocellulose products may impact production costs and market access, requiring companies to navigate complex regulatory landscapes [13, 14].

3. Types of nanocellulose

In general, depending on the source of cellulose, the production method, their size, and function, there are three types of nanocellulose with similar chemical composition and different morphology, namely nano-fibrillated cellulose (NFC), cellulose nanocrystals (CNCs), and bacterial nanocellulose (BNC) [15].

Nano-fibrillated celluloses are 1–100 nm in diameter and 500–2000 nm in length (depending on the type of raw material), and all are long, entangled, and flexible. Moreover, they consist of 100% cellulose in both amorphous and crystalline forms, and in comparison to CNCs, NFCs possess a greater surface area, higher aspect ratio, and longer length [16]. Also, among various types of nanocelluloses, NFCs have the simplest production method, owing to their avoidance of harsh chemical processes that degrade the molecular structure of cellulose. Generally, they are prepared by the physical separation of cellulose fibers, such as grinding, homogenization, and ultrasonication, or by chemical methods, including oxidation of wood raw material by 2,6,6-Tetramethylpiperidine-1-oxyl radical (TEMPO) under gentle stirring, and finally by a combination of two chemical and physical methods that include carboxymethylation and high-pressure homogenization [17].

Mostly, CNCs are obtained through acid hydrolysis of cellulose nanofibrils. Their diameter and length are 2–20 μm and 100–500 nm, respectively. Their structure is rod- or whisker-shaped and consists entirely of crystalline cellulose (around 54–88% crystallinity). CNCs exhibit an exceptionally high crystallinity [18]. The production of nanocellulose involves two fundamental steps. First, impurities of the raw material are eliminated using an alkaline (NaOH) solution or by a bleaching method. Subsequently, the raw material is heated in an acidic environment to eliminate the amorphous portion of cellulose, and after centrifugation and dialysis, CNC is obtained. It is worth noting that the type of raw material, type of acid, and hydrolysis temperature and time affect the dimensions of CNCs [19, 20].

Bacterial nanocelluloses have a twisted ribbon-like shape and length of 20–100 nm, which are chiefly obtained from *Gluconacetobacter xylinus, Acetobacter*, *Rhizobium*, *Agrobacterium*, *Aerobacter*, *Achromobacter*, *Azotobacter*, *Salmonella*, *Escherichia*, and *Sarcina* systems or cell-free systems [21, 22]. BNCs form a hydrogel at the water-air interface. Their synthesis involves the production of β-1,4-glucan chains within the bacterial cell excreted through the cell membrane as protofibrils. These protofibrils then crystallize to create ribbon-shaped microfibrils, leading to the formation of a pellicle. The physical and mechanical properties of BNCs are different depending on the type of bacterial strain, synthesis method (cell and cell-free systems), cultural conditions (carbon and nitrogen source), structural features, and arrangement of fibers [23].

4. Application of nanocellulose in the food industry

Nanocelluloses possess exceptional utility and distinctiveness across various industries, particularly in the food sector, attributed to their remarkable capacity for forming hydrogen bonds, expansive surface area, and aspect ratio compared to cellulose fibers. Additionally, their inexhaustible primary sources further enhance their significance. Different types of nanocellulose are used in the food industry as emulsifiers, fat substitutes, thickeners, and in food packaging. Nanocellulose also acts as a carrier agent in biological systems to carry nutrients like curcumin, vitamins, omega-3-fatty acids, minerals, amino acids, nutraceuticals, probiotics, polyphenols, antioxidants, micronutrients, enzymes, and essential oils [23].

4.1 Stabilizer and emulsifier

In the food industry, many natural substances such as gums (like xanthan, carrageenan, gellan, and locust bean gum) [24] and hydrophilic polysaccharides extracted from seaweed and microorganisms are used as natural stabilizers and emulsifiers [25]. The use of stabilizers in products like gravies, creams, sauces, frostings, additives, icings, salad dressings, foams, soups, puddings, dips, desserts, texture enhancers, thickeners, and frozen dairy products is highly prevalent [26]. However, subjects such as thinning, phase separation, and lack of phase stability over time are the main problems of the food industry. Hence, it is crucial to employ a cost-effective and easily extractable natural material that not only preserves the quality but also prevents any physical changes in the products [27]. Nanocellulose (NC), as a natural emulsifying and stabilizing ingredient, is of great interest in food products due to its unique rheological properties. Also, it can extend the shelf life and preserve the shape of frozen desserts. While increasing the amount of other stabilizers can improve shape retention, it may negatively impact the texture and flavor of the dessert. However, NC can prolong the structure retention time without compromising other desirable characteristics of the dessert [28]. In 2022, researchers successfully created medium internal phase oil-in-water (O/W) Pickering emulsions using bamboo shoots nanocellulose (BSNC) as a stabilizer. The nanocellulose extracted from bamboo shoots demonstrated promising potential as an emulsifier in the Pickering emulsions. Emulsions with a BSNC content of 0.5 wt% at a 5:5 oil-to-water volume ratio had smaller particle sizes of around 25. However, increasing the BSNC content resulted in larger droplet sizes and even demulsification. Subjecting the emulsions to twice shearing improved their physicochemical properties significantly by reducing droplet size. As the BSNC content increased, the apparent viscosity initially increased and then decreased, with all emulsions exhibiting elastic behaviors [29].

Another application of NC is the development of Pickering emulsions in various industries such as dairy, meat, and bakery. For example, in the dairy industry, it is a suitable option for replacing butter in baking cakes and replacing cream in preparing frozen yogurt and ice cream. In 2019, research was conducted to investigate the addition of NC to ice cream and its effect on the structure and performance of the frozen product. The results showed that NC has a direct effect on fat absorption. So, the presence of NC caused enhanced sensory attributes in samples containing less fat, and crystal growth, ice cream hardness, and rheology remained unchanged at subzero temperatures [30]. The surface tension between oil and water decreases by adding NC to different emulsions, which improves the stability of the emulsion system [31]. In this regard, the effect of adding NC as a stabilizer on meat sausage was investigated. The findings revealed that the addition of NC enhanced the binding capacity of fat and moisture, the sausage's hardness, elasticity, and chewability, resulting in reduced cooking loss and improved emulsion stability [32]. So, in general, Pickering emulsions stabilized with NCs have more stability and better compatibility than traditional surfactants.

4.2 Fat replacer

Today, with the progress of science, the world has moved toward consuming healthy and smart foods with low fat and sugar content and, as a result, suitable calories. Obesity and overweight are some of the basic problems among people of different societies, which have forced scientists to research the replacement of fattening factors, especially fat, with appropriate and natural substances. The use of NC in many foods, especially high-consumption and high-demand foods such as ice cream and processed meats, is one of the effective options for reducing the amount of fat and, as a result, reducing calories, maintaining the quality of the product's texture, and improving nutritional and health aspects. In this regard, Velásquez-Cock et al. [30] added NC to the ice cream formulation to produce ice cream with low fat and calories. The results showed that the presence of NC in low-fat ice creams caused a significant decrease in the melting rate of reduced-fat ice cream. A study conducted in 2022 aimed to develop reduced-fat mayonnaise formulations with 5, 15, and 30% fat content using varying concentrations of nanocellulose synthesized from palm-pressed fiber. The study also included a 20-day stability assessment. The results showed that the reduced-fat mayonnaise had smaller oil droplets closely packed together when viewed under a light microscope. However, significant oil droplet coalescence was observed during storage, which could lead to viscosity loss. These findings suggest that nanocellulose was effective as a fat mimetic during mayonnaise formulation, but its effectiveness during long-term storage is uncertain [33].

4.3 Dietary fiber

Nanocellulose, as a dietary fiber, has a significant impact on the general health of the body through three mechanisms: promoting satiety (effect on protein), regulating blood sugar (effect on glucose), and regulating blood lipids (impact on lipid and fat). Foods containing high protein are mainly used to control obesity and weight due to their high satiety effect; therefore, a more complex protein has a slower digestion rate, and as a result, the time to induce satiety in the body increases. In general, the speed of food digestion determines the availability of nutrients in the body, and the delay in emptying the stomach may evoke the satiety effect [34, 35]. The results of recent years have shown that NC, like anionic polysaccharides (pectin or carrageenan), effectively enhances digesta viscosity, impedes proteolysis, and decelerates gastric emptying through their interaction with protein solutions. NCs with a negative charge can decrease protein proteolysis in the stomach by attracting proteins with opposite charges through electrostatic forces. Consequently, this leads to stable satiety and reduced consumption of other foods (**Figure 1**) [36].

Another benefit of NC as a dietary fiber is its role in the treatment of obesity and diabetes. NC impedes glucose absorption in the body by slowing down the digestion and absorption of glucose. NC spontaneously combines with glucose-decomposing enzymes to form a hydrophobic complex, which changes the protein structure of the enzyme and disrupts the activity of α-amylase and amyl glucosidase (**Figure 2**) [37]. In 2019, scientists also investigated the effect of NC, extracted from bleached softwood pulp, on digestive viscosity, glucose digestion, and absorption. The findings indicated that concentrations exceeding 0.5 wt% substantially enhanced digesta viscosity, slowed hydrolysis, minimized glucose release, and delayed glucose diffusion [39]. Also, in 2023, to increase the value of waste chili stems and promote the recycling of green resources, cellulose was extracted from chili stems using a nitric acid-ethanol method, and nanocellulose was then prepared using a sulfuric acid hydrolysis method. The study demonstrated that biscuits with satisfactory overall quality could be made using 7% nanocellulose, resulting in a regular appearance and a relatively smooth surface. Mice experiments revealed that consuming biscuits containing nanocellulose reduced food intake significantly and inhibited weight gain in mice. Therefore, the research suggests that whole wheat biscuits with nanocellulose could be considered as a high-fiber food option [40].

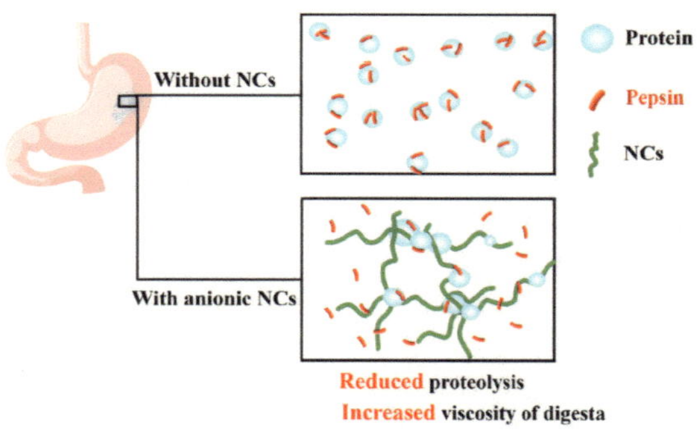

Figure 1.
Schematic of the effect of NC on protein digestion [36].

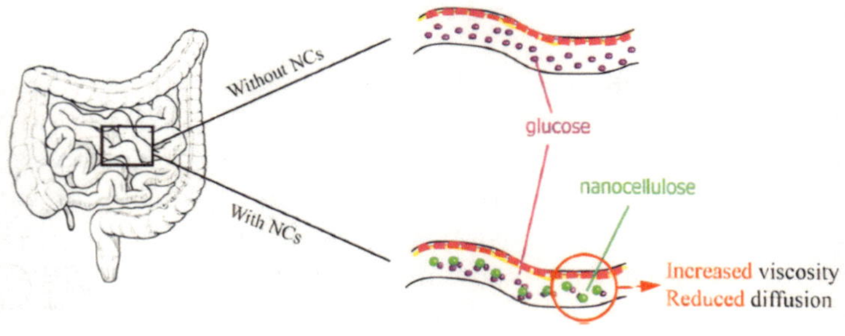

Figure 2.
Schematic of NC effect on glucose digestion [37, 38].

Another application of NC in the food industry is its use as an agent that disrupts the digestion and absorption of lipids in the body. The presence of lipids and fats in daily meals, including milk, ice cream, hamburgers, bread, and dairy products, plays a significant role in weight gain and blood pressure increase. Therefore, the production of targeted and intelligent food holds a key role in the future of the food industry. In general, NCs, with their absorption power and high water content, increase food's viscosity, which in turn reduces the food's caloric content and causes disorders in fat digestion. So, NCs effectively hinder the digestion and full absorption of lipids in the body through various fundamental mechanisms. These include (1) thickening or gelling gastrointestinal fluids to prevent digestive enzymes from reaching the fat; (2) combining with gastrointestinal constituents such as bile salt to form a precipitate; (3) reducing the surface area exposed to digestive enzymes by promoting flocculation of fat; and (4) forming a thin coating around the fat to reduce its exposure to digestive enzymes (**Figure 3**) [41]. To investigate the effect of NC on lipids, scientists simulated the conditions and environment of the digestive system. They found that NC reduces the activity of the lipase enzyme, increases the viscosity of intestinal digesta, and is more effective than cellulose in blocking the effect of bile salts and absorbing cholesterol in the body [42].

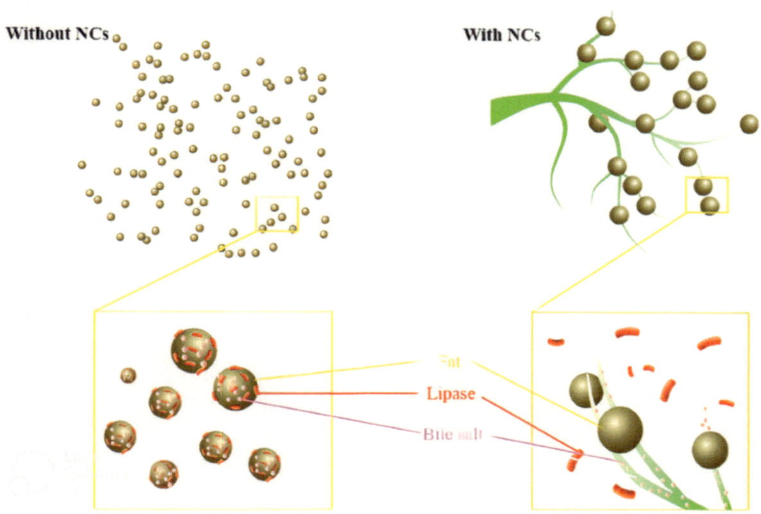

Figure 3.
Schematic of NC effect on fat digestion [41].

4.4 Retrogradation inhibitor in starchy food

Another nutritional application of NC is its use in the starchy food industry. Scientists have found that nanocellulose (especially CNC) can be used to produce healthy starchy foods and support weight loss efforts. A major issue in the starchy food industry is "retrogradation," which typically occurs during the cooling and storage stage of gelatinized starch, reducing the quality of the product [43]. With a large specific surface area and abundant hydroxyl groups, CNC effectively interacts with amylose through hydrogen bonding, inhibiting short-term retrogradation. Furthermore, the interaction between CNC and amylopectin inhibits long-term retrogradation (**Figure 4**) [44]. There have been extensive studies on the effect of NC on starch. For example, a 2017 study investigating the impact of NC on oxidized potato starch found that NC influenced enzyme activity and the amount of resistant starch. Notably, the addition of NC to gelatinized starch reduced the amount of quickly digestible and slowly digestible starch but increased the amount of resistant starch. Resistant starch is a type of starch that resists enzymatic digestion and cannot be digested in the small intestine. However, it can be fermented with volatile fatty acids in the human intestinal tract, specifically in the colon [45].

4.5 Food additive

Today, nanocelluloses, especially BNCs, due to small diameter, hydrophobicity, and water-holding capacity, are added to food and are used to reduce calories, increase and improve their absorption in the body, and improve nutritional properties. Among the applications of nanocellulose as a food additive, it can be mentioned to replace BNCs instead of flavonoids, cara gum, and other stabilizers to the ice cream formula to obtain a new type of ice cream. For example, in 2018, Guo et al. [46] showed that adding bacteria to ice cream increased its resistance to melting and could maintain its shape for at least an hour after being taken out of the freezer. Since BNCs are easily digested and absorbed by the digestive system, they are used as a quality improver based on meat. For example, in 2019, Guo et al. [47] demonstrated that BNC can effectively reduce moisture loss in chicken cake without altering the chemical

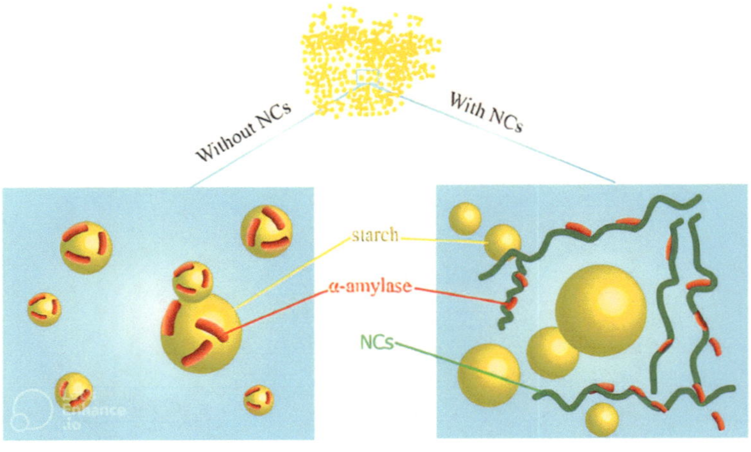

Figure 4.
Schematic of NC effect on starch digestion [37, 38].

structure or protein groups in the chicken. Another application of nanocellulose is the enrichment of beverages and dairy products in the market. For example, nanocelluloses can produce quality yogurt with a soft texture and higher stability [48].

4.6 Encapsulating agent

The food industry faces a significant challenge in preserving micronutrients, vitamins, minerals, and active components (such as probiotics) during production, preparation, transportation, and storage. Moreover, digestive enzymes often compromise these nutrients in the stomach and bile salts before reaching their intended destination in the body, primarily the intestine [49]. Therefore, food industry scientists are striving to protect nutrients from adverse factors and conditions and improve the performance of the delivery system using encapsulation techniques. Encapsulation is an active substance transfer system that uses natural polymer materials to safeguard nutrients, thereby enhancing the bioavailability and biological activity of substances. The polymer materials used in encapsulation should ideally be natural, readily available, inexpensive, non-toxic, free of pollution, easily modifiable, and have good biodegradability and biocompatibility [34]. Among the materials used for encapsulation, NC has drawn the attention of numerous researchers due to its distinctive biological properties, such as high safety, excellent biodegradability, and non-toxicity. A noteworthy study in this area has explored the use of alginate and NC in the microencapsulation of probiotics. The results of the study indicated that the inclusion of NC increases the compression strength of the produced hydrogels by more than 30% and diminishes the porosity induced by freeze-drying. Additionally, the introduction of 13% NC significantly extended the survival time of the probiotic strain (*Lactobacillus rhamnosus*) within the alginate microbeads [50]. In 2020, researchers encapsulated the probiotic *Lactobacillus rhamnosus* using a mixture of whey protein isolate, inulin, and NC (CNC). They found that NC significantly improved the resistance and survivability of the microcapsules under simulated digestive system conditions [51]. Also, Wang et al. [52] developed a soy protein isolate combined with CNC composite nanoparticles for the delivery of curcumin. The resulting complex composite nanoparticles demonstrated improved encapsulation efficiency and controlled release in simulated gastrointestinal conditions. CNC was found to be essential in the delivery system for the active compounds. In 2021, a compelling study was conducted on the encapsulation of ascorbic acid (Vitamin C) using NC. In this experiment, Vitamin C was encapsulated through an electrostatic interface with glycidyltrimethylammonium chloride-modified chitosan (GCS) and then cross-linking with phosphorylated CNC (PCNC) to produce Vitamin C-GCS-PCNC nanocapsules. The results of the investigation revealed that the inclusion of NC in the encapsulation formulation enhanced the efficiency, provided better control of the release action, and increased the antioxidant and antibacterial capabilities of vitamin C [53].

4.7 Food packaging materials

Packaging plays an essential role in protecting food from spoilage factors, including light, oxygen, heat, and bacteria. This makes it a vital component within the comprehensive system of the food industry. The use of natural and renewable materials in efficient packaging not only preserves food and extends its shelf life but also maintains the quality and nutritional value of the food until its expiration date, thereby helping to minimize waste in the sector. The prevalent use of plastic and its derivatives in the packaging industry, combined with increasing concern about environmental pollution and public health, has promoted scientists to explore natural and biodegradable materials for the production of suitable packaging. Among these materials, cellulose and its derivatives, especially NCs, have emerged as significant and essential candidates for the creation of biodegradable and intelligent packaging. One of the prominent attributes of NC that inspires scientists to employ it in the packaging industry includes its natural origin, high biodegradability, strength, resistance to oxygen and moisture penetration, and antibacterial properties. Generally, the exceptional oxygen barrier properties of NCs result from a combination of their high crystallinity and a network structure sustained by strong inter- and intramolecular hydrogen bonds [54]. Extensive research is being conducted on the application of NC in the packaging industry, particularly in intelligent packaging. For instance, in 2021, researchers manufactured nanocomposite films to regulate D-limonene permeability by utilizing nanocrystals and starch. Their study results indicated that the amount of NC, along with its appropriate aspect ratio, could autonomously manage D-limonene permeability [55]. In another study, Perumal, et al. [56] discovered that incorporating NC into the formulation of a biocomposite film (comprising clay and polyvinyl alcohol (PVA)) not only prolonged the storage life of mangoes but also enhanced the thermal and tensile properties of the films by increasing the amount of CNC filler up to 6 wt %. Currently, researchers are exploring the use of NC in intelligent packaging systems, leveraging its antibacterial and antioxidant properties. In 2020, researchers produced a probiotic nanocomposite film utilizing *Lactobacillus plantarum*, NC, and inulin. Their findings indicated that the fabricated probiotic packaging films exhibited potent antibacterial properties and effectively extended the shelf life of chicken fillets [57]. Another study produced composite films using polyhydroxybutyrate (PHB) and nanocellulose (CNC) via the solvent casting method. This study showed that, besides enhancing the barrier performance, mechanical strength, transparency, and other attributes of the composite film, CNC did not impact the inherent PHB [58].

Another beneficial effect of incorporating nanocellulose into packaging systems is enhancing the mechanical properties, such as high tensile strength, elongation, and flexural modulus [59]. For instance, a 2020 study observed a sevenfold increase in tensile strength and a sixfold increase in elongation at break when CNC was added to a PVA film [60]. Moreover, in research conducted by Chen et al. [61], it was found that integrating 0.5% stearic acid-modified microcrystalline cellulose/nanocellulose into cassava starch film significantly increased tensile strength by 484.5%.

5. Safety and regulatory aspects

Wood pulp and powdered cellulose are generally recognized as safe (GRAS) and are commonly used as raw materials in food-contact materials or food additives. As NC is one of the most widely used cellulose derivatives in different industries, evaluating its biological effects on the ecosystem and the human body solely based on its chemical characteristics is not sufficient. The interactions of NC with cells and other living organisms can be influenced by various factors, including its size, shape, aggregation characteristics, and other unknown factors [62]. One of the primary routes for NC to enter the body is through the skin and respiratory system. If the amount of NC in the skin and respiratory system exceeds permissible limits, it can cause skin toxicity and pulmonary inflammation [34, 63]. The use of commercial foodborne nanoparticles, such as titanium dioxide (TiO_2) and zinc oxide (ZnO), in the food industry can lead to complications like colitis [64], and this raises significant concerns among consumers and researchers about the safety of NC following ingestion through the gastrointestinal tract. Therefore, extensive studies have been conducted to investigate the effect of NC and related problems on food and the human body, using both *in vivo* and *in vitro* methods.

In 2010, the toxicity effect, including the *in vitro* cytotoxic and genotoxic properties of NC, obtained from birch pulp, with dimensions ranging between 20 and 60 nm to 100 and 350 nm, on human and animal cells was investigated. Additionally, the study examined the lethal effect of NC on human cells. Also, the effect of NC damage on DNA using a bacteriological assay (the Ames test) was also studied [65]. In 2011, another *in vitro* study focused on the toxicity, health, and environmental safety of NC with dimensions of 20–30 nm. The study evaluated its effect on human and mouse macrophages, its impact on cell survival, as well as the cytokinetic characteristics of cells [66]. Also, in recent years, extensive studies have been conducted on the oral toxicity of NCs using both *in vitro* cellular models and/or *in vivo* animal models. For instance, in 2019, the toxicological effects of ingested NC were investigated using both an *in vitro* intestinal epithelium model and an *in vivo* rat model [67]. In 2020, the impact of CNF on gut microbial health was studied by examining its cytotoxicity and *in vivo* toxicity [68], and in 2020, the hepatotoxicity of NC modified with oxalate ester in rats was investigated [69]. Furthermore, in 2021, the oral toxicity of carboxymethylated CNF was evaluated using *in vitro* cellular studies and an *in vivo* mice repeated gavage model [70]. All these studies consistently concluded that both original and functionalized NC showed no cytotoxicity. Additionally, it was observed that NC was non-toxic in an *in vivo* animal model when used at a reasonable concentration, with an upper limit of 3.5 wt%. Therefore, to gather clearer information regarding the safety effects of nanomaterials like NC, it is crucial to establish internationally standardized methods that guarantee their safety. Moreover, prior to the commercialization of food products utilizing new nanotechnologies, it is essential to proactively anticipate, understand, and effectively manage both the potential positive and negative consequences that may arise from consuming nanomaterials [62].

6. Conclusion and prospects

Cellulose, the most abundant polymer in nature, is the primary resource for the production of NC. The unique properties of NC, including its biocompatibility, biodegradability, non-toxicity, inexhaustible nature, nanoscale size, high aspect ratios, large surface area, liquid crystalline, and rheological properties, have made it an appealing choice for use in different industries such as food industry. Generally, NC has a greater affinity for oil than water. It can be used as a barrier to mix water and oil emulsions, making it a valuable stabilizing agent for emulsions and food products. This characteristic explains the use of NC in the food industry, particularly in foods such as sauces, soups, puddings, etc. Moreover, due to its ability to create stable emulsions with low-oil content, NC can contribute to the production of low-oil, low-calorie food products. NC can also serve as a suitable coating and carrier for bioactive compounds, such as probiotics, protecting them from the effects of digestive enzymes. Consequently, NC can play a crucial role in food enrichment. When utilized in packaging systems, NC enhances the mechanical and barrier properties of packaging films. As a result, food products are less exposed to spoilage factors, and their shelf life is increased. From the economic point of view, the use of nanocellulose in the industry due to its cheap source, low weight, and high strength, as well as enhanced product performance, can have a bright future, provided that it can overcome the upcoming obstacles, such as the expensive production method and the instability of the market. Also, there are still significant challenges faced during the application of NC in the food industry. These challenges include improving technology and reducing production costs, developing new methods for modifying the hydrophobicity of NC, further investigating its nutritional properties and impact on consumer health, and conducting more comprehensive research on the safety aspects of NC. As research and development continue to advance the production and utilization of nanocellulose, its economic viability is expected to further improve in the coming years.

Author details

Elham Asghari-Varzaneh and Hajar Shekarchizadeh*
Department of Food Science and Technology, College of Agriculture, Isfahan University of Technology, Isfahan, Iran

*Address all correspondence to: shekarchizadeh@iut.ac.ir

IntechOpen

© 2023 The Author(s). Licensee IntechOpen. This chapter is distributed under the terms of the Creative Commons Attribution License (http://creativecommons.org/licenses/by/3.0), which permits unrestricted use, distribution, and reproduction in any medium, provided the original work is properly cited.

References

[1] Liu Y, Ahmed S, Sameen DE, Wang Y, Lu R, Dai J, et al. A review of cellulose and its derivatives in biopolymer-based for food packaging application. Trends in Food Science & Technology. 2021;**112**:532-546. DOI: 10.1016/j.tifs.2021.04.016

[2] de Amorim JDP, de Souza KC, Duarte CR, da Silva Duarte I, de Assis Sales Ribeiro F, Silva GS, et al. Plant and bacterial nanocellulose: Production, properties and applications in medicine, food, cosmetics, electronics and engineering. A review. Environmental Chemistry Letters. 2020;**18**:851-869. DOI: 10.1007/s10311-020-00989-9

[3] He X, Lu W, Sun C, Khalesi H, Mata A, Andaleeb R, et al. Cellulose and cellulose derivatives: Different colloidal states and food-related applications. Carbohydrate Polymers. 2021;**255**:117334. DOI: 10.1016/j.carbpol.2020.117334

[4] Zainal SH, Mohd NH, Suhaili N, Anuar FH, Lazim AM, Othaman R. Preparation of cellulose-based hydrogel: A review. Journal of Materials Research and Technology. 2021;**10**:935-952. DOI: 10.1016/j.jmrt.2020.12.012

[5] Trache D, Hussin MH, Chuin CTH, Sabar S, Fazita MN, Taiwo OF, et al. Microcrystalline cellulose: Isolation, characterization and bio-composites application—A review. International Journal of Biological Macromolecules. 2016;**93**:789-804. DOI: 10.1016/j.ijbiomac.2016.09.056

[6] Lavoine N, Desloges I, Dufresne A, Bras J. Microfibrillated cellulose–Its barrier properties and applications in cellulosic materials: A review. Carbohydrate Polymers. 2012;**90**(2):735-764. DOI: 10.1016/j.carbpol.2012.05.026

[7] Abdel-Hakim A, Mouradb RM. Nanocellulose and its polymer composites: Preparation, characterization, and applications. Russian Chemical Reviews. 2023;**92**:4. DOI: 10.57634/RCR5076

[8] Kaur P, Sharma N, Munagala M, Rajkhowa R, Aallardyce B, Shastri Y, et al. Nanocellulose: Resources, physio-chemical properties, current uses and future applications. Frontiers in Nanotechnology. 2021;**3**:747329. DOI: 10.3389/fnano.2021.747329

[9] Sharma A, Thakur M, Bhattacharya M, Mandal T, Goswami S. Commercial application of cellulose nano-composites–A review. Biotechnology Reports. 2019;**21**:e00316. DOI: 10.1016/j.btre.2019.e00316

[10] Klemm D, Cranston ED, Fischer D, Gama M, Kedzior SA, Kralisch D, et al. Nanocellulose as a natural source for groundbreaking applications in materials science: Today's state. Materials Today. 2018;**21**(7):720-748. DOI: 10.1016/j.mattod.2018.02.001

[11] Mateo S, Peinado S, Morillas-Gutiérrez F, La Rubia MD, Moya AJ. Nanocellulose from agricultural wastes: Products and applications—A review. Processes. 2021;**9**(9):1594. DOI: 10.3390/pr9091594

[12] Hisseine OA, Wilson W, Sorelli L, Tolnai B, Tagnit-Hamou A. Nanocellulose for improved concrete performance: A macro-to-micro investigation for disclosing the effects of cellulose filaments on strength of cement systems. Construction and Building Materials. 2019;**206**:84-96. DOI: 10.1016/j.conbuildmat.2019.02.042

[13] Penloglou G, Basna A, Pavlou A, Kiparissides C. Techno-economic

considerations on nanocellulose's future progress: A short review. Processes. 2023;**11**(8):2312. DOI: 10.3390/pr11082312

[14] Clauser NM, Felissia FF, Area MC, Vallejos ME. Technological and economic barriers of industrial-scale production of nanocellulose. In: Green Nanomaterials for Industrial Applications. Amesterdam, The Netherlands: Elsevier; 2022. pp. 21-39. DOI: 10.1016/B978-0-12-823296-5.00015-0

[15] Qi Y, Guo Y, Liza AA, Yang G, Sipponen MH, Guo J, et al. Nanocellulose: A review on preparation routes and applications in functional materials. Cellulose. 2023;**30**(7):4115-4147. DOI: 10.1007/s10570-023-05169-w

[16] Yang X, Li L, Nishiyama Y, Reid MS, Berglund LA. Processing strategy for reduced energy demand of nanostructured CNF/clay composites with tailored interfaces. Carbohydrate Polymers. 2023;**312**:120788. DOI: 10.1016/j.carbpol.2023.120788

[17] Zhang F, Shen R, Li N, Yang X, Lin D. Nanocellulose: An amazing nanomaterial with diverse applications in food science. Carbohydrate Polymers. 2022;**304**:120497. DOI: 10.1016/j.carbpol.2022.120497

[18] Xie J, Liu S. A review of hydrophobic nanocellulose and its applications. Paper and Biomaterials. 2021;**6**(2):35-42. DOI: 10.12103/j.issn.2096-2355.2021.02.004

[19] Mateo S, Peinado S, Morillas-Gutiérrez F, La Rubia MD, Moya AJ. Nanocellulose from agricultural wastes: Products and applications—A review. Processes. 2021;**9**(9):1594. DOI: 10.3390/pr9091594

[20] Dhali K, Ghasemlou M, Daver F, Cass P, Adhikari B. A review of nanocellulose as a new material towards environmental sustainability. Science of the Total Environment. 2021;**775**:145871. DOI: 10.1016/j.scitotenv.2021.145871

[21] Ullah MW, Ul-Islam M, Khan S, Kim Y, Park JK. Structural and physico-mechanical characterization of bio-cellulose produced by a cell-free system. Carbohydrate Polymers. 2016;**136**:908-916. DOI: 10.1016/j.carbpol.2015.10.010

[22] Tahir D, Karim MRA, Hu H, Naseem S, Rehan M, Ahmad M, et al. Sources, chemical functionalization, and commercial applications of nanocellulose and nanocellulose-based composites: A review. Polymers. 2022;**14**(21):4468. DOI: 10.3390/polym14214468

[23] Perumal AB, Nambiar RB, Moses J, Anandharamakrishnan C. Nanocellulose: Recent trends and applications in the food industry. Food Hydrocolloids. 2022;**127**:107484. DOI: 10.1016/j.foodhyd.2022.107484

[24] Mu R, Hong X, Ni Y, Li Y, Pang J, Wang Q, et al. Recent trends and applications of cellulose nanocrystals in food industry. Trends in Food Science & Technology. 2019;**93**:136-144. DOI: 10.1016/j.tifs.2019.09.013

[25] Pang B, Liu H, Zhang K. Recent progress on Pickering emulsions stabilized by polysaccharides-based micro/nanoparticles. Advances in Colloid and Interface Science. 2021;**296**:102522. DOI: 10.1016/j.cis.2021.102522

[26] Zanchetta G, Rocchi E, Piazza L. Seeing is believing: Coupling between liquid crystalline ordering and rheological behavior in cellulose nanocrystals suspensions. Chemical Engineering Transactions. 2017;**57**:1933-1938. DOI: 10.3303/CET1757323

[27] Sharma G, Sharma S, Kumar A, Ala'a H, Naushad M, Ghfar AA, et al. Guar

gum and its composites as potential materials for diverse applications: A review. Carbohydrate Polymers. 2018;**199**:534-545. DOI: 10.1016/j.carbpol.2018.07.053

[28] Winuprasith T, Suphantharika M. Properties and stability of oil-in-water emulsions stabilized by microfibrillated cellulose from mangosteen rind. Food Hydrocolloids. 2015;**43**:690-699. DOI: 10.1016/j.foodhyd.2005.06.002

[29] Lei Y, Zhang X, Li J, Chen Y, Liang H, Li Y, et al. Nanocellulose from bamboo shoots as perfect Pickering stabilizer: Effect of the emulsification process on the interfacial and emulsifying properties. Food Bioscience. 2022;**46**:101596. DOI: 10.1016/j.fbio.2022.101596

[30] Velásquez-Cock J, Serpa A, Vélez L, Gañán P, Hoyos CG, Castro C, et al. Influence of cellulose nanofibrils on the structural elements of ice cream. Food Hydrocolloids. 2019;**87**:204-213. DOI: 10.1016/j.foodhyd.2018.07.035

[31] Costa ALR, Gomes A, Tibolla H, Menegalli FC, Cunha RL. Cellulose nanofibers from banana peels as a Pickering emulsifier: High-energy emulsification processes. Carbohydrate Polymers. 2018;**194**:122-131. DOI: 10.1016/j.carbpol.2018.04.001

[32] Qi W, Wu J, Shu Y, Wang H, Rao W, Xu H-N, et al. Microstructure and physiochemical properties of meat sausages based on nanocellulose-stabilized emulsions. International Journal of Biological Macromolecules. 2020;**152**:567-575. DOI: 10.1016/j.ijbiomac.2020.02.285

[33] Lee ZJ, Tong SC, Tang TK, Lee YY. Palm-based cellulose nanofiber isolated from mechano-chemical processing as sustainable rheological modifier in reduced fat mayonnaise. Journal of Food Science. 2022;**87**(8):3542-3561. DOI: 10.1111/1750-3841.16250

[34] Zhang K, Wang W, Zhao K, Ma Y, Wang Y, Li Y. Recent development in foodborne nanocellulose: Preparation, properties, and applications in food industry. Food Bioscience. 2021;**44**:101410. DOI: 10.1016/j.fbio.2021.101410

[35] Bashir O, Rashid S, Masoodi N, Khan SA, Majid I, Malik M. Nanocellulose: Chemistry, preparation, and applications in the food industry. In: Industrial Applications of Nanocellulose and Its Nanocomposites. Cambridge: Woodhead Publishing, Elsevier; 2022. pp. 155-177. DOI: 10.1016/B978-0-323-89909-3.00008-0

[36] Liu L, Kong F. Influence of nanocellulose on in vitro digestion of whey protein isolate. Carbohydrate Polymers. 2019;**210**:399-411. DOI: 10.1016/j.carbpol.2019.01.071

[37] Nsor-Atindana J, Goff HD, Saqib MN, Chen M, Liu W, Ma J, et al. Inhibition of α-amylase and amyloglucosidase by nanocrystalline cellulose and spectroscopic analysis of their binding interaction mechanism. Food Hydrocolloids. 2019;**90**:341-352. DOI: 10.1016/j.foodhyd.2018.12.031

[38] Liu L, Kerr WL, Kong F, Dee DR, Lin M. Influence of nano-fibrillated cellulose (NFC) on starch digestion and glucose absorption. Carbohydrate Polymers. 2018;**196**:146-153. DOI: 10.1016/j.carbpol.2018.04.116

[39] Liu L, Kong F. In vitro investigation of the influence of nano-cellulose on starch and milk digestion and mineral adsorption. International Journal of Biological Macromolecules. 2019;**137**:1278-1285. DOI: 10.1016/j.ijbiomac.2019.06.194

[40] Ma Y, Chai X, Bao H, Huang Y, Dong W. Study on nanocellulose isolated from waste chilli stems processing as dietary fiber in biscuits. PLoS One. 2023;**18**(1):e0281142. DOI: 10.1371/journal.pone.0281142

[41] DeLoid GM, Sohal IS, Lorente LR, Molina RM, Pyrgiotakis G, Stevanovic A, et al. Reducing intestinal digestion and absorption of fat using a nature-derived biopolymer: Interference of triglyceride hydrolysis by nanocellulose. ACS Nano. 2018;**12**(7):6469-6479. DOI: 10.1021/acsnano.8b03074

[42] Liu L, Kong F. In vitro investigation of the influence of nano-fibrillated cellulose on lipid digestion and absorption. International Journal of Biological Macromolecules. 2019;**139**:361-366. DOI: 10.1016/j.ijbiomac.2019.07.189

[43] Chen L, Ren F, Zhang Z, Tong Q, Rashed MM. Effect of pullulan on the short-term and long-term retrogradation of rice starch. Carbohydrate Polymers. 2015;**115**:415-421. DOI: 10.1016/j.carbpol.2014.09.006

[44] Cui S, Li M, Zhang S, Liu J, Sun Q, Xiong L. Physicochemical properties of maize and sweet potato starches in the presence of cellulose nanocrystals. Food Hydrocolloids. 2018;**77**:220-227. DOI: 10.1016/j.foodhyd.2017.09.037

[45] Ji N, Liu C, Li M, Sun Q, Xiong L. Interaction of cellulose nanocrystals and amylase: Its influence on enzyme activity and resistant starch content. Food Chemistry. 2018;**245**:481-487. DOI: 10.1016/j.foodchem.2017.10.130

[46] Guo Y, Zhang X, Hao W, Xie Y, Chen L, Li Z, et al. Nano-bacterial cellulose/soy protein isolate complex gel as fat substitutes in ice cream model. Carbohydrate Polymers. 2018;**198**:620-630. DOI: 10.1016/j.carbpol.2018.06.078

[47] Guo Y, Wu J, Li T, Zhang S, Wang S, Feng X, et al. Effects of bacterial cellulose on chicken quality and protein properties. Journal of Northwest A & F University-Natural Science Edition. 2019;**47**(3):129-137

[48] Lu Q, Yu X, Yagoub AEA, Wahia H, Zhou C. Application and challenge of nanocellulose in the food industry. Food Bioscience. 2021;**43**:101285. DOI: 10.1016/j.fbio.2021.101285

[49] Zhang R, Zhang Z, Zhang H, Decker EA, McClements DJ. Influence of emulsifier type on gastrointestinal fate of oil-in-water emulsions containing anionic dietary fiber (pectin). Food Hydrocolloids. 2015;**45**:175-185. DOI: 10.1016/j.foodhyd.2014.11.020

[50] Huq T, Fraschini C, Khan A, Riedl B, Bouchard J, Lacroix M. Alginate based nanocomposite for microencapsulation of probiotic: Effect of cellulose nanocrystal (CNC) and lecithin. Carbohydrate Polymers. 2017;**168**:61-69. DOI: 10.1016/j.carbpol.2017.03.032

[51] Maleki O, Khaledabad MA, Amiri S, Asl AK, Makouie S. Microencapsulation of lactobacillus rhamnosus ATCC 7469 in whey protein isolate-crystalline nanocellulose-inulin composite enhanced gastrointestinal survivability. LWT- Food Science and Technology. 2020;**126**:109224. DOI: 10.1016/j.lwt.2020.109224

[52] Wang S, Lu Y, Ouyang X-k, Ling J. Fabrication of soy protein isolate/cellulose nanocrystal composite nanoparticles for curcumin delivery. International Journal of Biological Macromolecules. 2020;**165**:1468-1474. DOI: 10.1016/j.ijbiomac.2020.10.046

[53] Baek J, Ramasamy M, Willis NC, Kim DS, Anderson WA, Tam KC. Encapsulation and controlled release of vitamin C in modified cellulose nanocrystal/chitosan nanocapsules. Current Research in Food Science. 2021;**4**:215-223. DOI: 10.1016/j.crfs.2021.03.010

[54] Le Normand M, Moriana R, Ek M. The bark biorefinery: A side-stream of the forest industry converted into nanocomposites with high oxygen-barrier properties. Cellulose. 2014;**21**:4583-4594. DOI: 10.1007/s10570-014-0423-z

[55] Liu S, Li X, Chen L, Li L, Li B, Zhu J. Tunable d-limonene permeability in starch-based nanocomposite films reinforced by cellulose nanocrystals. Journal of Agricultural and Food Chemistry. 2018;**66**(4):979-987. DOI: 10.1021/acs.jafc.7b05457

[56] Perumal AB, Sellamuthu PS, Nambiar RB, Sadiku ER, Phiri G, Jayaramudu J. Effects of multiscale rice straw (*Oryza sativa*) as reinforcing filler in montmorillonite-polyvinyl alcohol biocomposite packaging film for enhancing the storability of postharvest mango fruit (*Mangifera indica* L.). Applied Clay Science. 2018;**158**:1-10. DOI: 10.1016/j.clay.2018.03.008

[57] Zabihollahi N, Alizadeh A, Almasi H, Hanifian S, Hamishekar H. Development and characterization of carboxymethyl cellulose based probiotic nanocomposite film containing cellulose nanofiber and inulin for chicken fillet shelf life extension. International Journal of Biological Macromolecules. 2020;**160**:409-417. DOI: 10.1016/j.ijbiomac.2020.05.066

[58] Seoane IT, Cerrutti P, Vázquez A, Manfredi LB, Cyras VP. Polyhydroxybutyrate-based nanocomposites with cellulose nanocrystals and bacterial cellulose. Journal of Polymers and the Environment. 2017;**25**:586-598. DOI: 10.1007/s10924-016-0838-8

[59] Hassan MM, Tucker N, Le Guen MJ. Thermal, mechanical and viscoelastic properties of citric acid-crosslinked starch/cellulose composite foams. Carbohydrate Polymers. 2020;**230**:115675. DOI: 10.1016/j.carbpol.2019.115675

[60] Jancy S, Shruthy R, Preetha R. Fabrication of packaging film reinforced with cellulose nanoparticles synthesised from jack fruit non-edible part using response surface methodology. International Journal of Biological Macromolecules. 2020;**142**:63-72. DOI: 10.1016/j.ijbiomac.2019.09.066

[61] Chen Q, Shi Y, Chen G, Cai M. Enhanced mechanical and hydrophobic properties of composite cassava starch films with stearic acid modified MCC (microcrystalline cellulose)/NCC (nanocellulose) as strength agent. International Journal of Biological Macromolecules. 2020;**142**:846-854. DOI: 10.1016/j.ijbiomac.2019.10.024

[62] Serpa A, Velásquez-Cock J, Gañán P, Castro C, Vélez L, Zuluaga R. Vegetable nanocellulose in food science: A review. Food Hydrocolloids. 2016;**57**:178-186. DOI: 10.1016/j.foodhyd.2016.01.023

[63] Stoudmann N, Schmutz M, Hirsch C, Nowack B, Som C. Human hazard potential of nanocellulose: Quantitative insights from the literature. Nanotoxicology. 2020;**14**(9):1241-1257. DOI: 10.1080/17435390.2020.1814440

[64] Cao X, Han Y, Gu M, Du H, Song M, Zhu X, et al. Foodborne titanium dioxide nanoparticles induce stronger adverse effects in obese mice than non-obese

mice: Gut microbiota dysbiosis, colonic inflammation, and proteome alterations. Small. 2020;**16**(36):2001858. DOI: 10.1002/smll.202001858

[65] Pitkänen M, Honkalampi U, Von Wright A, Sneck A, Hentze H-P, Sievänen J, et al. Nanofibrillar cellulose: In vitro study of cytotoxic and genotoxic properties. In: 2010 TAPPI International Conference on Nanotechnology for the Forest Product Industry; 27-29. Espoo, Finland: TAPPI Press; Semtember 2010. pp. 246-261. Available from: http://www.tappi.org/content/events/10nano/papers/5.4.pdf

[66] Vartiainen J, Pöhler T, Sirola K, Pylkkänen L, Alenius H, Hokkinen J, et al. Health and environmental safety aspects of friction grinding and spray drying of microfibrillated cellulose. Cellulose. 2011;**18**:775-786. DOI: 10.1007/s10570-011-9501-7

[67] DeLoid GM, Cao X, Molina RM, Silva DI, Bhattacharya K, Ng KW, et al. Toxicological effects of ingested nanocellulose in in vitro intestinal epithelium and in vivo rat models. Environmental Science: Nano. 2019;**6**(7):2105-2115. DOI: 10.1039/C9EN00184K

[68] Khare S, DeLoid GM, Molina RM, Gokulan K, Couvillion SP, Bloodsworth KJ, et al. Effects of ingested nanocellulose on intestinal microbiota and homeostasis in Wistar Han rats. NanoImpact. 2020;**18**:100216. DOI: 10.1016/j.impact.2020.100216

[69] Otuechere C, Adewuyi A, Adebayo O, Ebigwei I. In vivo hepatotoxicity of chemically modified nanocellulose in rats. Human & Experimental Toxicology. 2020;**39**(2):212-223. DOI: 10.1177/0960327119881672

[70] Zhang K, Zhang H, Wang W. Toxicological studies and some functional properties of carboxymethylated cellulose nanofibrils as potential food ingredient. International Journal of Biological Macromolecules. 2021;**190**:887-893. DOI: 10.1016/j.ijbiomac.2021.09.055

Chapter 5

Bacterial Nanocellulose: Methods, Properties, and Biomedical Applications

Haiyong Ao and Xiaowei Xun

Abstract

Unlike plant and wood-origin cellulose, bacterial nanocellulose (BNC) produced by bacteria exhibits the highest purity and natural nanofiber morphology, attracting increasing interest from many researchers and industrial sectors. It has numerous unique features including the biomimetic nanoscale three-dimensional (3D) network, high water holding capacity, and moldability in different shapes, accepted wet strength, outstanding gas permeability, and good biocompatibility, which makes the BNC show great potential in a wide variety of biomedical applications. Extensive research has verified the feasibility of application in wound dressing, bone/cartilage tissue regeneration, vascular tissue engineering, and so on. This chapter focuses on the production and properties of BNC, the fabrication of BNC-based biomaterials, and the biomedical applications of BNC.

Keywords: bacterial nanocellulose, fabrication techniques, properties, biomaterials, biomedical applications

1. Introduction

Bacterial nanocellulose (BNC) as a type of extracellular polysaccharide polymer synthesized by bacteria has unique physical and chemical properties, including high purity (free from lignin and hemicellulose) and crystallinity (84–89%), high degree of polymerization, good mechanical properties, high water retention, and three-dimensional (3D) nanofibrous structure [1, 2]. All these features make BNC attractive for material scientists and engineers.

Importantly, BNC is environment-friendly due to its nontoxic, nonimmunogenic, biocompatible, biodegradable, and renewable nature [3]. The nanofibrillar network of BNC is similar to the structure of collagen in the native extracellular matrix (ECM), and the 3D uniform and interconnected pores facilitate cell infiltration and nutrient and waste exchange [4, 5]. As a promising natural biomaterial, BC has been extensively utilized in various biomedical fields for wound healing, bone, cartilage, and blood vessel engineering [6, 7].

During the past two decades, many studies have been dedicated to developing various BC-based biomaterials. The 3D porous structure and easily modified surface

IntechOpen

(abundant OH groups) of BNC are beneficial for introducing into BC varieties of reinforcement substances including biomolecules, nanoparticles, and polymers to acquire new materials with highly desirable properties, which can lead to the formation of BC-based functional materials for biomedical applications [3]. Moreover, BC-based scaffolds have been specifically designed to mimic the 3D structures of native tissues to support and provide the microenvironments required for cell adhesion, proliferation, migration, and differentiation [8]. These extensive efforts have accelerated the development of BNC-based biomaterial.

In this chapter, we briefly introduced the biosynthesis and properties of BNC. Furthermore, the strategies for fabricating the BNC biomaterial and applications in various biomedical fields were summarized.

2. Biosynthesis and properties of BNC

2.1 Biosynthesis and production of BNC

The bacterial synthesis of cellulose was first reported by Brown as early as 1886 [9]. Brown found that a white gel-like thin layer was formed on the surface of the culture medium after incubating *Bacterium xylinum*' in a static state. Chemical analysis determined that the main component of the white gel is cellulose and does not contain hemicellulose and lignin, and further research found that the diameter of this cellulose is at the nanoscale.

Gluconacetobacter genus, as one of the most efficient BNC producers, is widely used in the biosynthesis of BNC [10, 11]. The general process to produce BNC is schematically described in **Figure 1A**. The biosynthesis process of BNC is mainly divided into four steps (**Figure 1B**): (1) glucose is converted into glucose-6-phosphate, (2) glucose-6-phosphate is converted into glucose-1-phosphate by phosphoglucomutase, (3) uridine diphosphate glucose (UDP-glucose) pyrophosphorylase converts glucose-1-phosphate to UDP-glucose, (4) the UDP-glucose is converted to cellulose by the action of cellulose synthase (**Figure 1C**) [13, 14]. The self-assembly and crystallization of BNC nanofibrils are regulated by the cell itself [15]. The synthesized chains are sequentially gathered in the culture medium at a rate of up to 200,000 glucose molecules per second and form protofibrils or sub-elementary fibrils, which are 2–4 nm in diameter. Each sub-elementary protofibril is composed of 12–16 glucan chains, where the glucan chains are aligned and stacked into ordered nanostructures called the microfibrils. The micro- and macrofibrils and loose bundles finally form 30–100 nm ribbon-like structures. The ribbons are highly porous and have 3D fibrous network.

For many years, many researches have been focused on BNC production. Various culture methods have been used by controlling the fermentation conditions to increase the yield of BNC or obtain BNC with different characteristics [16]. Under static culture conditions, bacteria need to float on the surface of the medium to obtain enough oxygen, and cellulose microfibrils are extruded from the bacteria and synthesize a tight BNC pellicle. The BNC formed under static conditions usually has high crystallinity and tensile strength. In contrast, a fluffy, spherical, or irregular lump BNC was obtained *via* the agitating culture method. Although the agitating culture has faster cell growth rate than static conditions because of more oxygen filling in the

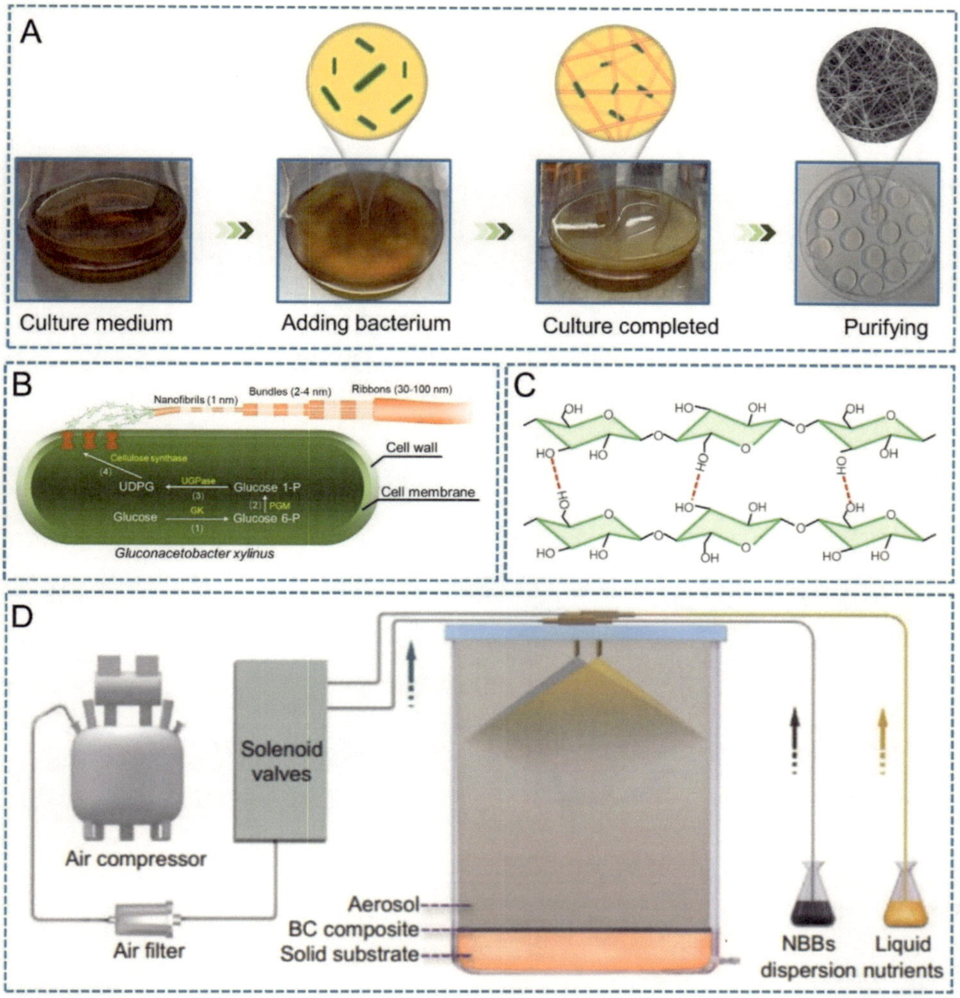

Figure 1.
(A) A schematic illustration of BNC production. (B) Schematic illustration of biochemistry of cellulose synthesis in a bacterial cell and the extracellular transport of cellulose chains and the formation of highly ordered structures. (C) Chemical structure of BNC. (D) Schematic illustration of the aerosol-assisted biosynthesis of BNC [12].

culture medium, the production of BNC decreased, which perhaps due to the uniform aeration of cultures induced cells to grow intensively instead of the polymer synthesis. Based on this, the bioreactors have been developed that can produce BNC pellicles at higher yields under static conditions (**Figure 1D**) [12].

2.2 Properties of BNC

Compared to vegetal cellulose, its unique chemical composition and physical structure endows BNC with numerous unparalleled physical, chemical, and biological properties (**Figure 2**). BCN is an attractive candidate for widespread applications in various fields, especially in applications related to biomedical.

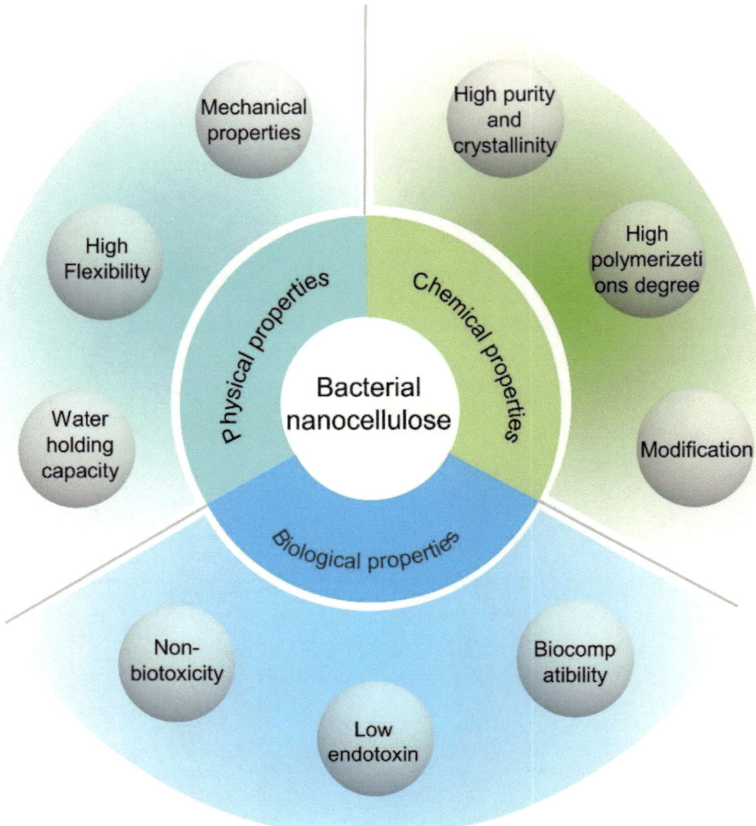

Figure 2.
A schematic representation of properties of BNC.

2.2.1 Physical properties

BNC has natural 3D porous network structure, which is composed of nanofibers and exhibits enormous mechanical properties. The Young's modulus of BNC sheets up to 15 GPa, which is much greater than plant cellulose and several synthetic fibers [17]. The excellent mechanical properties make BNC to be used for blood vessels and bone tissue engineering. The water holding capacity of BNC is about 100 times its dry weight, which is attributed to the high surface area and pore volume of BNC that can intercept more water [18]. In addition, the average diameter of BNC nanofibers is 1.5 nm, showing higher surface area and flexibility. These fascinating physical properties make BNC become desirable wound dressing material.

2.2.2 Chemical properties

BNC is composed of linear homopolysaccharides conjugated by β-D-glucose units linked by 1,4-β-glycosidic linkages and has considerably higher crystallinity (80–90%) and degree of polymerization (up to 8000), which means that pure BNC can be obtained with simple processing. The abundant hydroxyl groups and high-level hydrogen bonds on the surface of BNC nanofibers allow for manipulation in their loading of functional molecules. Although BNC exhibits significant intrinsic characteristics, it is still necessary to develop more modified properties of BNC to meet the requirements of the required biomedical applications.

2.2.3 Biological properties

BNC has wide applications in biomedical engineering because of its excellent biological properties. Héctor and coworkers confirmed the nontoxicity and extremely low bacterial endotoxin of BNC nanofibers by using *in vitro* and *in vivo* tests [19]. In addition, the good biocompatibility of BNC is due to its peculiar 3D nanofibrous network structure that supports cell penetration and proliferation. Helenius et al. subcutaneously implanted the BNC in rats to systematically evaluate the biocompatibility of BNC [20]. The results showed no inflammation around the implants and the fibroblasts infiltrated BNC, which indicates the good biocompatibility of BNC and has the potential to be used as a scaffold in tissue engineering.

3. Fabrication of BNC biomaterials

Although the above excellent inherent characteristics make BNC have broad application prospects in the field of biomedical, pure BC possesses certain restrictions that limit its application due to the lack of unique biological functions. In recent years, several approaches have been developed to improve the physicochemical properties and function of BNC, including producing BNC composites, surface medication, and changes in porosity. These efforts have significantly improved the surface properties and biological functions of BNC and tremendously expanded its application in various tissue engineering fields.

3.1 Surface modifications

3.1.1 Physical modifications

The BNC nanofibers were physical modifications by using bioactive materials, which not only preserves the unique properties but also enhances the biological functions to meet the requirements of biomedical applications. To improve the biocompatibility of BNC, Luo et al. selected a new type of coating, chondroitin sulfate (CS) modified with gelatin (Gel), to modify BNC scaffold [21]. The coating on the surface of the BNC nanofibers is linked by hydrogen bonding. The results of *in vitro* cell studies indicated that the CS/Gel coatings significantly promote cell proliferation, adhesion, differentiation, and ingrowth into scaffolds. Nanomaterials can also be used for surface modification of BNC. He et al. developed a Cu^{2+} loaded phase-transited lysozyme (PTL) nanocoating for surface modification of BNC. The coating endowed the BNC with inhibiting bacterial growth and inflammation and simultaneously induced vascularization, collagen deposition, and reepithelialization of wounds promising dressings for healing infected wound, which was considered a promising dressing for treating infected wounds. In addition, polymer composite nanoparticles are also used for surface modification of fibers. Ma et al. developed a homogeneous AgNP-loaded polydopamine (PDA)/polyethyleneimine (PEI) coating on the surface of BNC nanofibers, which exhibited excellent antibacterial activities and cytocompatibility [22].

3.1.2 Chemical modifications

The only surface functional group on the surface of BNC nanofibers is the hydroxyl group, which is not suitable for cell proliferation [23]. Nevertheless, the hydroxyl groups on the surface of BNC nanofibers are beneficial for surface chemical modifications; they can be treated with various chemical reagents tomodify their chemical structure and incorporate additional functionalities. Therefore, the surface chemical modifications of BNC are required for biomedical applications. Oxidation is an important reaction for adding new functional groups to BNC, which creates infinite possibilities for the functionalization of BNC nanofibers. TEMPO (2,2,6,6-Tetramethylpiperidinyloxy or 2,2,6,6-Tetramethylpiperidine-1-oxyl radical)-mediated oxidization is a simple and effective chemical reaction, in which C_6 of a hydroglucose units selectively gain an anionic carboxyl group [24]. A new bone repair composite scaffold (CS/OBC/nHAP) was constructed by evenly dispersing the *in situ* crystalline nano-hydroxyapatite (nHAP) in oxidized bacterial cellulose (OBC) and chitosan (CS) scaffolds [7]. Shahriari-Khalaji et al. achieved high carboxylate content by optimizing the TEMPO-mediated oxidation of BNC and then covalently bonded the ε-poly-L-lysine (PLL) with oxidized BNC to develop an O-BNC-based functional wound dressing [25]. Furthermore, Xie et al. prepared biofunctional group-modified bacterial cellulose (DCBC) by carboxymethylation and selective oxidation to achieve the perfect compound of cellulose and chitosan [26]. The obtained BNC-based dressing can effectively inhibit bacterial proliferation in wounds and kill the bacteria.

3.2 BNC-based composites

3.2.1 BNC nanocomposites

To improve the mechanical and biological properties of BNC biomaterials, various BNC-based nanocomposites were prepared by incorporating different kinds of nanomaterials, including carbon-based nanoparticles (NPs) [27], metal/metal oxide NPs [28], and other inorganic NPs [29]. The simplest and most convenient method for fabricated BNC nanocomposites is the mechanical mixing method. Yang et al. prepared the BNC/Ti3C2Tx nanocomposites using mechanical mixing. Although this strategy allows for sufficient mixing of nanomaterials with BNC nanofibers, it damages the intrinsically continuous 3D structure of BNC. To maintain the intrinsically continuous 3D structure of BC, Luo et al. developed a cost-effective, scalable, and efficient approach, membrane-liquid interface (MLI) culture, to prepare BNC-based nanocomposite [30]. The MLI culture technology is schematically illustrated in **Figure 3A** [31]. First, a layer of BNC film (around 3 mm in thickness) was grown as a substrate using a conventional static culture method. Then, a nanomaterials-dispersed medium was sprayed onto the BNC substrate followed by the culturing of BNC on the substrate–medium interface. After the sprayed medium was completely consumed off by the BNC growth, another layer of the medium was sprayed. This process was repeated until the designed thickness was achieved. It is noted that the thickness of nanocomposites prepared using the MLI method can be accurately controlled by the spray cycles of the medium. In addition, the sample shape and dimension can be facially designed by the geometry of the containers used for the preparation. This method has been widely used to prepare various nanocomposites (**Figure 3B**) [12].

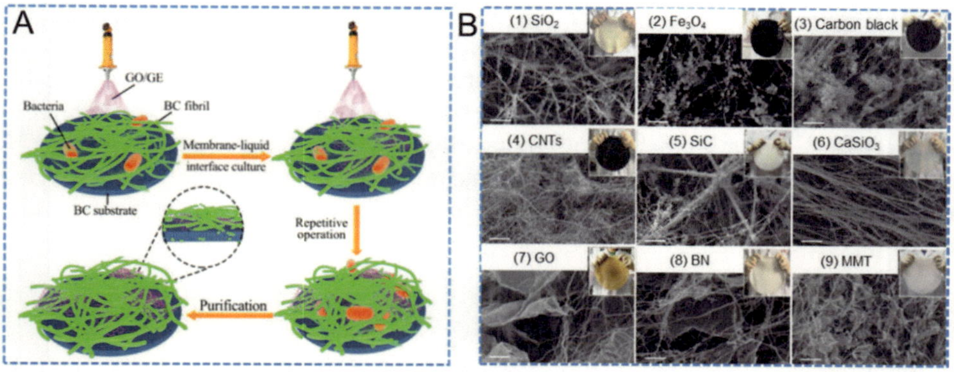

Figure 3.
(A) Schematic illustration of the preparation of BNC nanocomposites by using membrane-liquid interface culture technology [31]. (B) SEM and photographs of BC-based nanocomposites were prepared by using membrane-liquid interface culture technology [12].

3.2.2 BNC-based biocomposites

Compared with other natural biopolymers, pristine BNC lacks cytocompatibility and important biological functions. The abundant −OH groups on the surface of BNC provide binding stable sites for biopolymers in biocomposites. Therefore, various techniques including *in situ* addition, solution impregnation, and chemical cross-linking were used to combine various bioactive substances to obtain the BNC-based biocomposites with enhanced physicomechanical, antimicrobial, and biocompatible properties. In order to enhance the antibacterial properties and cytocompatibility of BNC, Zhou et al. composited the collagen I (Col-I) and the antibacterial agent hydroxypropyltrimethyl ammonium chloride chitosan (HACC) into the BNC 3D network structure by a novel membrane-liquid interface (MLI) culture (**Figure 4A** and **B**) [32]. The introduction of HACC and Col-I makes BNC have outstanding antibacterial properties and improved cytocompatibility to promote NIH3T3 cell and HUVEC proliferation and

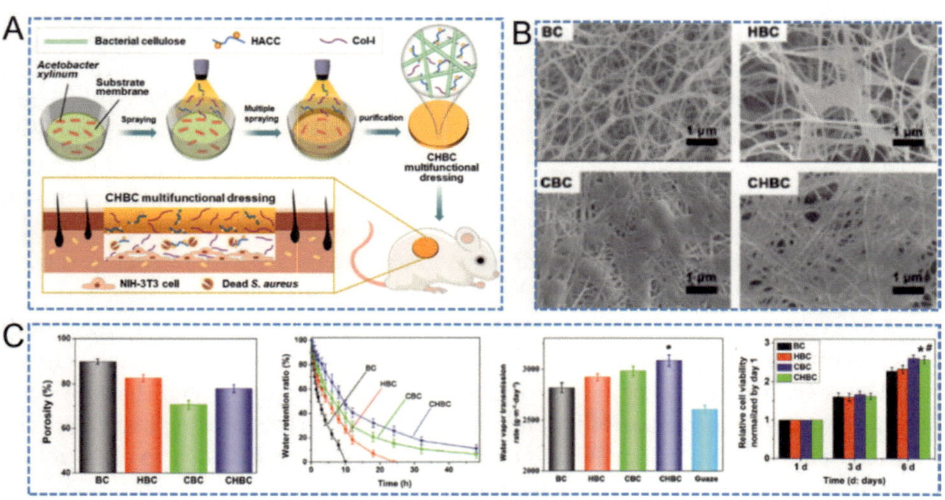

Figure 4.
(A) Schematic illustration of the preparation of BNC-based biocomposites. (B) SEM images of BNC-based biocomposites. (C) The characteristice of BNC-based biocomposites [32].

spread (**Figure 4C**). The BNC-(polypyrrole) Ppy composites were prepared to mimic the natural myocardial microenvironment by *in situ* polymerization [33]. The composites were flexible and still maintained 3D network structure and displayed electrical conductivities in the range of native cardiac tissue. Wan et al. reported extracellular matrix (ECM)-mimetic scaffolds by conjugating electrospun cellulose acetate (CA) submicrofibers with BNC nanofibers *via* a facile and scalable dispersion freeze-drying process [34]. It is found that the composites have a 3D porous network structure and improved cell behavior.

3.3 3D porous BNC scaffold

As a promising biomaterial, BNC has some intrinsic disadvantages when applied in tissue engineering; its dense nanofibrous network (the pore sizes of pristine BNC are only approximately 0.02–10 μm) markedly limits cell migration and 3D tissue regeneration. To date, different methods have been reported for the fabrication of 3D porous BNC scaffolds. The paraffin microparticles, potato starch, agarose microparticles, and gelatin microspheres as porogens have also been used to fabricate 3D porous BNC scaffolds by using the *in situ* porogen technique [35–38]. Cui et al. described the utilization of gelatin microspheres in BNC production for porosity enhancement, and the observed porosity depended on the diameter of microspheres (**Figure 5A**) [36]. However, a nonuniform pore structure would be formed due to the bacteria moving to the air/medium interface during the culture process of the *in situ* porogen impregnation technique.

Laser patterning as an efficient method is used to prepare 3D porous BNC scaffold. Laser treatment has universality, and the obtained scaffolds do not contain pollutants and chemical cross-linking agents, thereby having better biocompatibility. The pore structure prepared by laser is parallel microchannels. Yang et al. fabricated the nano-submicrofibrous cellulose scaffolds with microchannels by laser-aided punching (**Figure 5B**) [39]. The cell study found that the presence of microchannels favors cell proliferation and migration at an optimum microchannel size. However, laser-aided punching prepared microchannels can lead to cell leakage from the scaffold, making it difficult to achieve 3D culture.

Freeze-drying is the most common technique for preparing a 3D porous BNC scaffold, which can obtain the BNC sponges with higher porosity and specific surface. However, the pore size of freeze-dried BNC sheets is unable to precisely regulate, and the pore structure is instability. Therefore, improvements need to be made to address these issues. Xun et al. first reported an improved strategy to fabricating 3D macroporous BNC scaffolds with controllable pore size by freeze-drying and cross-linking the mechanically disintegrating shortcut BNC nanofiber suspensions (**Figure 5C**) [40]. The pore sizes of the MP-BNC scaffolds were controlled by adjusting the concentration of BNC in the suspensions. The fabrication process was facile, scalable, and effective in controlling the pore structure. The cross-linked BNC scaffolds exhibited excellent compression properties and shape recovery ability compared to the original BNC. Moreover, the results of *in vitro* and *in vivo* studies demonstrated that the scaffolds had excellent biocompatibility and were effective in regenerating cartilage tissue.

Additive manufacturing or 3D printing is an emerging technology to prepare 3D porous scaffolds through rapid prototyping. In recent years, it has been widely used in personalized customization of tissue engineering scaffolds. Li et al. used the gelatin methacrylate (GelMA)/BNC bioink formulations to develop heterogeneous tissue-engineered skin (HTS) containing layers of fibroblast networks with larger

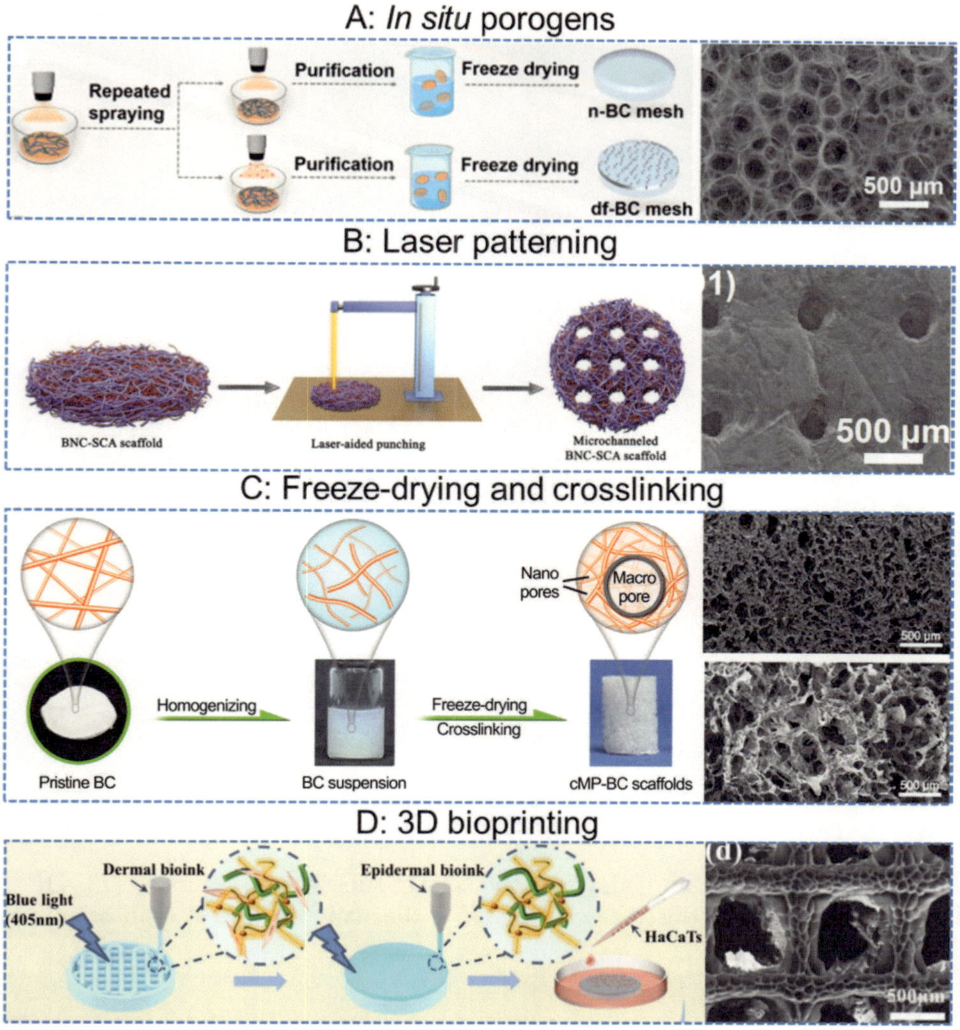

Figure 5.
Different 3D porous BNC scaffolds were fabricated by using in situ *porogen [34], laser patterning [39], freeze-drying and cross-linking [40], and 3D bioprinting [41].*

pores, basal layers with smaller pores, and multilayered keratinocytes (**Figure 5D**) [41]. The results revealed that the 10%GelMA/0.3%BNC bioink was better to bioprint dermis due to its high printability and cell-friendly sparse microenvironment. The approaches developed in the above-described studies and their findings suggest that BNC has great potential to be printed into 3D microstructures for the development of scaffolds and medical devices for various biomedical applications.

4. Biomedical applications of BNC

In the past two decades, BNC has aroused great attention to fundamental and scientific research for biomedical applications due to its unique properties [42]. Functionalized BNC and its composites have been applied in several medical applications, such as wound dressing, bone and cartilage tissue regeneration, and the

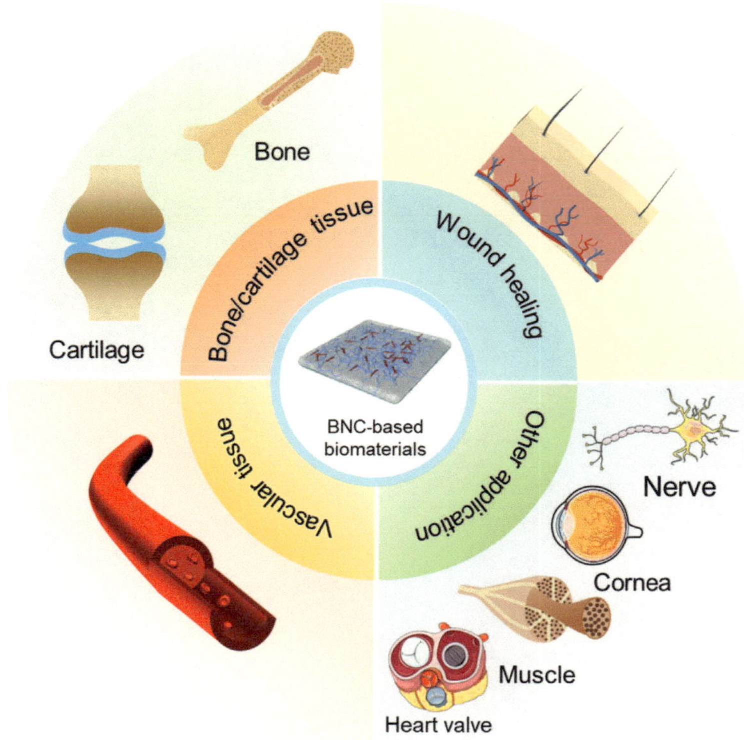

Figure 6.
A schematic representation of the potential biomedical applications of BNC biomaterials.

development of artificial organs and blood vessels' substitutes (**Figure 6**) [23, 43]. This section focuses on the research progress of the main biomedical applications of BNC in recent years.

4.1 Wound healing

The skin is the main protective barrier of the human body and has many functions, such as controlling body temperature and maintaining a balance of electrolytes and water [44]. However, large-scale skin damage caused by trauma, ulcers, and burns is difficult to heal on its own and requires human intervention to promote wound healing [45, 46]. Wound dressings have been developed to prevent infection and dehydration of wound, reduce inflammatory responses, and promote wound healing. BNC is one of the most promising wound dressings due to its ultrafine 3D network structure, excellent gas permeability, high water absorbency, and favorable biocompatibility [47]. Furthermore, BNC-based dressings can also prevent bacterial invasion, absorb excess exudates, and retain moisture in wounds to promote the growth of granulation.

In the healing process, it is important to prevent bacterial infection on the wound site. However, BNC does not have intrinsic bactericidal properties to eliminate the colonized bacteria. Therefore, the initial research focus was on antibacterial BNC-based dressings. Hence, various promising antibacterial agents are introduced into BNC, such as metals and their oxides, antimicrobial peptides, and biological and synthetic polymers (chitosan) [48]. Ao et al. fabricated the quaternized chitosan (HACC)/BNC antibacterial wound dressing by using *in situ* synthesis possessed, and the obtained dressing had favorable antibacterial properties [49]. To further improve the antibacterial properties of BNC dressings, silver nanoparticles (AgNPs) were introduced into the BNC structure [22]. The results of the standard plate count assay indicated that the antibacterial BNC dressing showed antibacterial rates of over 99.9% against *Escherichia coli* and *Staphylococcus aureus*. The *in vivo* assay demonstrated that the antibacterial BNC dressing could inhibit infection and inflammation and accelerate wound healing within 12 days compared with BNC. To reduce the cytotoxicity of sliver-based dressing and promote wound healing, a novel multifunctional sliver nanowires (AgNWs)/collagen I (Col I)/BNC dressing was constructed *via* compositing AgNWs into BNC and adsorbing Col I [50]. The antibacterial assay demonstrated that the multifunctional AgNWs/Col I/BNC dressing could kill *S. ausreus* and *E. coli* colonizing on the surface of material due to the silver ions released. Importantly, this dressing had lower cytotoxicity and promoted collagen deposition, hair follicle growth, neovascularization, and wound healing. A novel copper ion (Cu^{2+})-loaded BNC-based antibacterial wound dressing was prepared *via* codeposition of polydopamine (PDA) and Cu^{2+} [51]. The *in vivo* study revealed that the dressing can eliminate *S. aureus* infections and inflammatory response and promote collagen deposition, capillary angiogenesis, and wound healing.

In addition, BNC composites containing nanomaterials, biopolymers, antioxidants, antibiotics, and blood clotting agents have been fabricated to improve the wound healing of BNC-based dressings. Cai et al. fabricated a composite adhesive organo hydrogel by introducing BNC and platelet-rich plasma (PRP) into a poly-N-(tris[hydroxymethyl]methyl)acrylamide (THMA)/N-acryloyl aspartic acid (AASP) hybrid gel network infiltrated with glycerol/water binary solvent [52]. The PRP-loaded organo hydrogel has good tissue adhesion properties and releases a variety of growth factors to accelerate the wound healing process through collagen deposition and angiogenesis. Shen et al. developed an aggregation-induced emission (AIE) molecule BITT-composited BNC for wound healing [53]. The BNC-BITT composites retained the advantages of biocompatible of BNC and displayed photodynamic and photothermal synergistic antibacterial effects under irradiation of a 660 nm laser, which endowed the dressings with excellent wound healing performance in a mouse full-thickness skin wound model infected by multidrug-resistant bacteria. In response to hemostasis and repair of irregular and deep skin wounds, an injectable aldehyde BNC/polydopamine (DBNC/PDA) photothermal cryogel was prepared by oxidation polymerization method [6]. The PDA enhances the photothermal properties of DBNC/PDA cryogel to kill most bacteria and provides wound protection under near-infrared light. Otherwise, the DBNC/PDA low temperature gel has rapid hemostatic effect in the face of irregular and deep skin wounds. It is worth noting that the nanoenzyme with excellent peroxidase (POD) activity has been used to prepare BNC-based wound dressing. Zhang et al. introduced the metal-organic frameworks (MOF)-based nanocatalysts loaded with glucose oxidase (GOx) into the BNC-reinforced hydrogel for the treatment of diabetic foot ulcers [54]. The designed nanoenzyme could effectively catalyze the decomposition of glucose and *in situ* generate •OH for bacteria killing. In addition, this nanoenzyme-based hydrogel exhibited excellent hemostatic properties owing to the enhanced absorption capacity.

4.2 Bone/cartilage tissue regeneration

Another area of potential exciting application for BNC biomaterial is bone/cartilage tissue engineering due to its biomimetic ECM properties, excellent mechanical properties, and highly porous structure [7, 55, 56]. Biosynthetic BNC has some intrinsic disadvantages (the dense nanofibrous network) that markedly limit its applications in tissue engineering. However, the *in situ* biosynthetic BNC could serve as a surface coating for Ti implants to improve their biological function [57, 58]. Liu et al. reported a metal ions-containing BNC coating for functional Ti implant by *in situ* biosynthesis on the surface of Ti with complex shapes [59]. The results of *in vitro* and *in vivo* experiments confirmed that the functional BNC coating on the Ti can integrate the operative crevices and promote osteogenesis. To optimize the pore structure of BNC to meet the requirements of bone and cartilage tissue, Xun et al. developed a 3D macroporous BNC scaffold by freeze-drying and cross-linking the BNC shortcut nanofibers [40]. After the 3D macroporous BNC scaffolds were implanted into nude mice subcutaneously for 8 weeks, the neocartilage tissue with native cartilage appearance and abundant cartilage-specific extracellular matrix deposition was successfully regenerated. To construct the BNC scaffolds with structurally and biochemically biomimetic cartilage tissue microenvironment, the 3D hierarchical porous BNC/decellularized cartilage extracellular matrix (DCECM) scaffold was fabricated by freeze-drying technique after EDC/NHS chemical crosslinking [60]. The *in vitro* and *in vivo* tests indicated that the BC/DCECM scaffolds achieved satisfactory neocartilage tissue regeneration with superior original shape fidelity, exterior natural cartilage-like appearance, and histologically cartilage-specific lacuna formation and ECM deposition. Ling et al. fabricated a hierarchically porous SF/BC/MXene (FSCM) scaffold with ~20.0 μm macropore and nanofibrillar wall, which has excellent bone defect repair ability [61]. In addition, the BNC can also be used for the osteochondral repair. Lou et al. designed a bilayer structure osteochondral scaffold with a dense γ-Polyglutamic acid/carboxymethyl chitosan/BNC (PGA/CMCS/BNC) hydrogel cartilage layer and a porous nano HA-containing PGA/CMCS/BNC hydrogel osteogenic layer [62]. The *in vivo* experiments indicated that the scaffold with bioactive ions had a much better effect on the repair of osteochondral defects.

4.3 Blood vessels

BNC is a possible material to use for artificial blood vessels for small- or large-sized vascular grafts due to its good mechanical strength (a burst pressure of up to 880 mmHg), blood biocompatibility, and moldability [63]. The bioreactor to produce tubular BNC was developed to prepared BNC-based vascular grafts with excellent cytocompatibility and hemocompatibility [64, 65]. A tubular BNC graft with greater mechanical strength and thinner walls was obtained by mercerization; this technology made fewer platelets adhere to the luminal surface and promoted the proliferation of endothelial cells [63]. Mimicking the morphological structure of native blood vessels is critical for the development of vascular grafts. Vascular grafts with BNC nanofibers and submicrofibrous cellulose acetate (CA) were fabricated to mimic the morphological structure of native blood vessels. Regulating the content of BNC can reduce the thrombosis potential of stents and enhance endothelialization.

4.4 Other biomedical application

In addition to the above biomedical applications, BNC has been applied in other biomedical fields, such as nerve repair [66, 67] and muscle [33], corneal [68, 69], urethral [70], and intervertebral disc [71]. These researches will be sure of certain theoretical value and practical significance to the biomedical application of BNC.

5. Conclusion

In this chapter, we provide an overview of the fabrication and biomedical application of BNC-based biomaterials. First, the biosynthesis of BNC in biology, chemistry, and physics is introduced, and the properties of BNC are summarized. Furthermore, we introduce and discuss the various techniques to fabricate BNC-based composites and 3D porous BNC scaffolds to enhance the mechanical and biological properties. Due to the rapid development of the abovementioned technologies, BNC has a broad range of applications in biomedicine, including wound dressing, bone/cartilage tissue regeneration, vascular tissue engineering, and so on.

Acknowledgements

This work was supported by National Natural Science Foundation of China (grant nos. 82160355), the Science and Technology Research Project of Jiangxi Education Department (grant nos. GJJ2200657), and Natural Science Foundation of Jiangxi Province (grant nos. 20212ACB214002).

Conflict of interest

The authors declare no conflict of interest.

Author details

Haiyong Ao* and Xiaowei Xun
School of Materials Science and Engineering, East China Jiaotong University, Nanchang, PR China

*Address all correspondence to: aohyong@126.com

IntechOpen

© 2024 The Author(s). Licensee IntechOpen. This chapter is distributed under the terms of the Creative Commons Attribution License (http://creativecommons.org/licenses/by/3.0), which permits unrestricted use, distribution, and reproduction in any medium, provided the original work is properly cited.

References

[1] Rajwade JM, Paknikar KM, Kumbhar JV. Applications of bacterial cellulose and its composites in biomedicine. Applied Microbiology and Biotechnology. 2015;**99**:2491-2511

[2] Huang Y, Zhu C, Yang J, et al. Recent advances in bacterial cellulose. Cellulose. 2013;**21**:1-30

[3] Mbituyimana B, Liu L, Ye W, et al. Bacterial cellulose-based composites for biomedical and cosmetic applications: Research progress and existing products. Carbohydrate Polymers. 2021;**273**:118565

[4] Huang K, Liu W, Wei W, et al. Photothermal hydrogel encapsulating intelligently bacteria-capturing bio-mof for infectious wound healing. ACS Nano. 2022;**16**:19491-19508

[5] Torres F, Commeaux S, Troncoso O. Biocompatibility of bacterial cellulose based biomaterials. Journal of Functional Biomaterials. 2012;**3**:864-878

[6] Cao S, Zhang K, Li Q, et al. Injectable and photothermal antibacterial bacterial cellulose cryogel for rapid hemostasis and repair of irregular and deep skin wounds. Carbohydrate Polymers. 2023;**320**:121239

[7] Cao S, Li Q, Zhang S, et al. Oxidized bacterial cellulose reinforced nanocomposite scaffolds for bone repair. Colloids and Surfaces B: Biointerfaces. 2022;**211**:112316

[8] Khan S, Ul-Islam M, Ullah MW, et al. Fabrication strategies and biomedical applications of three-dimensional bacterial cellulose-based scaffolds: A review. International Journal of Biological Macromolecules. 2022;**209**:9-30

[9] Brown A. XLIII.—On an acetic ferment which forms cellulose. Journal of the Chemical Society, Transactions. 1886;**49**:432-439

[10] Zhang H, Xu X, Chen X, et al. Complete genome sequence of the cellulose-producing strain Komagataeibacter nataicola RZS01. Scientific Reports. 2017;**7**:4431

[11] Zhang H, Ye C, Xu N, et al. Reconstruction of a genome-scale metabolic network of komagataeibacter nataicola rzs01 for cellulose production. Scientific Reports. 2017;**7**:7911

[12] Guan QF, Han ZM, Luo TT, et al. A general aerosol-assisted biosynthesis of functional bulk nanocomposites. National Science Review. 2019;**6**:64-73

[13] Penttilä PA, Imai T, Capron M, et al. Multimethod approach to understand the assembly of cellulose fibrils in the biosynthesis of bacterial cellulose. Cellulose. 2018;**25**:2771-2783

[14] Manan S, Ullah MW, Ul-Islam M, et al. Bacterial cellulose: Molecular regulation of biosynthesis, supramolecular assembly, and tailored structural and functional properties. Progress in Materials Science. 2022;**129**:100972

[15] Geiger O, López-Lara IM, Sohlenkamp C. Phosphatidylcholine biosynthesis and function in bacteria. Biochimica et Biophysica Acta (BBA) - Molecular and Cell Biology of Lipids. 2013;**1831**:503-513

[16] Esa F, Tasirin SM, Rahman NA. Overview of bacterial cellulose production and application. Agriculture and Agricultural Science Procedia. 2014;**2**:113-119

[17] Yamanaka S, Watanabe K, Kitamura N, et al. The structure and mechanical properties of sheets prepared from bacterial cellulose. Journal of Materials Science. 1989;**24**:3141-3145

[18] Ul-Islam M, Khan T, Park JK. Water holding and release properties of bacterial cellulose obtained by in situ and ex situ modification. Carbohydrate Polymers. 2012;**88**:596-603

[19] Martinez Avila H, Schwarz S, Feldmann EM, et al. Biocompatibility evaluation of densified bacterial nanocellulose hydrogel as an implant material for auricular cartilage regeneration. Applied Microbiology and Biotechnology. 2014;**98**:7423-7435

[20] Helenius G, Bäckdahl H, Bodin A, et al. In vivo biocompatibility of bacterial cellulose. Journal of Biomedical Materials Research. 2006;**76A**:431-438

[21] Luo H, Yin C, Zhong B, et al. Modifying porous bacterial cellulose with chondroitin sulfate/gelatin for improved biocompatibility. Fibers and Polymers. 2023;**24**:975-984

[22] Ma L, Jiang W, Xun X, et al. Homogeneous silver nanoparticle loaded polydopamine/polyethyleneimine-coated bacterial cellulose nanofibers for wound dressing. International Journal of Biological Macromolecules. 2023;**246**:125658

[23] Gregory DA, Tripathi L, Fricker ATR, et al. Bacterial cellulose: A smart biomaterial with diverse applications. Materials Science and Engineering: R: Reports. 2021;**145**:100623

[24] Iwamoto S, Kai W, Isogai T, et al. Comparison study of TEMPO-analogous compounds on oxidation efficiency of wood cellulose for preparation of cellulose nanofibrils. Polymer Degradation and Stability. 2010;**95**:1394-1398

[25] Shahriari-Khalaji M, Li G, Liu L, et al. A poly-l-lysine-bonded TEMPO-oxidized bacterial nanocellulose-based antibacterial dressing for infected wound treatment. Carbohydrate Polymers. 2022;**287**:119266

[26] Xie Y, Qiao K, Yue L, et al. A self-crosslinking, double-functional group modified bacterial cellulose gel used for antibacterial and healing of infected wound. Bioactive Materials. 2022;**17**:248-260

[27] Yoon SH, Jin H-J, Kook M-C, et al. Electrically conductive bacterial cellulose by incorporation of carbon nanotubes. Biomacromolecules. 2006;**7**:1280-1284

[28] Foresti ML, Vázquez A, Boury B. Applications of bacterial cellulose as precursor of carbon and composites with metal oxide, metal sulfide and metal nanoparticles: A review of recent advances. Carbohydrate Polymers. 2017;**157**:447-467

[29] Celik KB, Cengiz EC, Sar T, et al. In-situ wrapping of tin oxide nanoparticles by bacterial cellulose derived carbon nanofibers and its application as freestanding interlayer in lithium sulfide based lithium-sulfur batteries. Journal of Colloid and Interface Science. 2018;**530**:137-145

[30] Luo H, Xiong P, Xie J, et al. Uniformly dispersed freestanding carbon nanofiber/graphene electrodes made by a scalable biological method for high-performance flexible supercapacitors. Advanced Functional Materials. 2018;**28**:1803075

[31] Luo H, Ao H, Peng M, et al. Effect of highly dispersed graphene and graphene oxide in 3D nanofibrous bacterial

cellulose scaffold on cell responses: A comparative study. Materials Chemistry and Physics. 2019;**235**:121774

[32] Zhou C, Yang Z, Xun X, et al. De novo strategy with engineering a multifunctional bacterial cellulose-based dressing for rapid healing of infected wounds. Bioactive Materials. 2022;**13**:212-222

[33] Srinivasan SY, Cler M, Zapata-Arteaga O, et al. Conductive bacterial nanocellulose-polypyrrole patches promote cardiomyocyte differentiation. ACS Applied Bio Materials. 2023;**6**:2860-2874

[34] Wan Y, Cui T, Zhang Q, et al. Submicrofiber-incorporated 3d bacterial cellulose nanofibrous scaffolds with enhanced cell performance. Macromolecular Materials and Engineering. 2018;**303**(11):1800316

[35] Zaborowska M, Bodin A, Bäckdahl H, et al. Microporous bacterial cellulose as a potential scaffold for bone regeneration. Acta Biomaterialia. 2010;**6**:2540-2547

[36] Cui T, Yu F, Zhang Q, et al. Double-layered bacterial cellulose mesh for hernia repair. Colloid and Interface Science Communications. 2021;**44**:100496

[37] Yang J, Lv X, Chen S, et al. In situ fabrication of a microporous bacterial cellulose/potato starch composite scaffold with enhanced cell compatibility. Cellulose. 2014;**21**:1823-1835

[38] Yin N, Stilwell MD, Santos TMA, et al. Agarose particle-templated porous bacterial cellulose and its application in cartilage growth in vitro. Acta Biomaterialia. 2015;**12**:129-138

[39] Yang Z, Yu F, Gan D, et al. Microchannels in nano-submicro-fibrous cellulose scaffolds favor cell ingrowth. Cellulose. 2021;**28**:9645-9659

[40] Xun X, Li Y, Zhu X, et al. Fabrication of robust, shape recoverable, macroporous bacterial cellulose scaffolds for cartilage tissue engineering. Macromoleculer Bioscience. 2021;**21**:e2100167

[41] Li M, Sun L, Liu Z, et al. 3D bioprinting of heterogeneous tissue-engineered skin containing human dermal fibroblasts and keratinocytes. Biomaterials Science. 2023;**11**:2461-2477

[42] Popa L, Ghica MV, Tudoroiu EE, et al. Bacterial cellulose-a remarkable polymer as a source for biomaterials tailoring. Materials. 2022;**15**:1054

[43] Yang Y, Lu Y-T, Zeng K, et al. Recent progress on cellulose-based ionic compounds for biomaterials. Advanced Materials. 2021;**33**:2000717

[44] Pereira RF, Bártolo PJ. Traditional therapies for skin wound healing. Advances in Wound Care. 2016;**5**:208-229

[45] Liang Y, He J, Guo B. Functional hydrogels as wound dressing to enhance wound healing. ACS Nano. 2021;**15**:12687-12722

[46] Dong R, Guo B. Smart wound dressings for wound healing. Nano Today. 2021;**41**:101290

[47] Ahmed J, Gultekinoglu M, Edirisinghe M. Bacterial cellulose micro-nano fibres for wound healing applications. Biotechnology Advances. 2020;**41**:107549

[48] Ahmad H. Celluloses as support materials for antibacterial agents: A review. Cellulose. 2021;**28**:2715-2761

[49] Ao H, Jiang W, Nie Y, et al. Engineering quaternized chitosan in

the 3D bacterial cellulose structure for antibacterial wound dressings. Polymer Testing. 2020;**86**:106490

[50] Yang Z, Ma L, Han X, et al. A facile, biosynthetic design strategy for high-performance multifunctional bacterial cellulose-based dressing. Composites Part B: Engineering. 2022;**238**:109945

[51] Yang Z, Feng F, Jiang W, et al. Designment of polydopamine/bacterial cellulose incorporating copper (II) sulfate as an antibacterial wound dressing. Biomaterials Advances. 2022;**134**:112591

[52] Cai C, Zhu H, Chen Y, et al. Platelet-rich plasma composite organohydrogel with water-locking and anti-freezing to accelerate wound healing. Advanced Healthcare Materials. 2023;**12**:2301477

[53] Shen Z, Zhu W, Huang Y, et al. Visual multifunctional aggregation-induced emission-based bacterial cellulose for killing of multidrug-resistant bacteria. Advanced Healthcare Materials. 2023;**12**:2300045

[54] Zhang S, Ding F, Liu Y, et al. Glucose-responsive biomimetic nanoreactor in bacterial cellulose hydrogel for antibacterial and hemostatic therapies. Carbohydrate Polymers. 2022;**292**:119615

[55] Gu M, Fan S, Zhou G, et al. Effects of dynamic mechanical stimulations on the regeneration of in vitro and in vivo cartilage tissue based on silk fibroin scaffold. Composites Part B: Engineering. 2022;**235**:109764

[56] Yang X, Huang J, Chen C, et al. Biomimetic design of double-sided functionalized silver nanoparticle/bacterial cellulose/hydroxyapatite hydrogel mesh for temporary cranioplasty. ACS Applied Materials & Interfaces. 2023;**15**:10506-10519

[57] Dydak K, Junka A, Szymczyk P, et al. Development and biological evaluation of Ti6Al7Nb scaffold implants coated with gentamycin-saturated bacterial cellulose biomaterial. PLoS One. 2018;**13**:e0205205

[58] Zhao J, Tong H, Kirillova A, et al. A synthetic hydrogel composite with a strength and wear resistance greater than cartilage. Advanced Functional Materials. 2022;**32**(41):2205662

[59] Liu X, Wang D, Wang S, et al. Promoting osseointegration by in situ biosynthesis of metal ion-loaded bacterial cellulose coating on titanium surface. Carbohydrate Polymers. 2022;**297**:120022

[60] Li Y, Xun X, Xu Y, et al. Hierarchical porous bacterial cellulose scaffolds with natural biomimetic nanofibrous structure and a cartilage tissue-specific microenvironment for cartilage regeneration and repair. Carbohydrate Polymers. 2022;**276**:118790

[61] Ling T, Zha X, Zhou K, et al. A facile strategy toward hierarchically porous composite scaffold for osteosarcoma ablation and massive bone defect repair. Composites Part B: Engineering. 2022;**234**:109660

[62] Luo M, Chen M, Bai J, et al. A bionic composite hydrogel with dual regulatory functions for the osteochondral repair. Colloids and Surfaces. B, Biointerfaces. 2022;**219**:112821

[63] Zhang Q, He S, Zhu X, et al. Heparinization and hybridization of electrospun tubular graft for improved endothelialization and anticoagulation. Materials Science and Engineering: C. 2021;**122**:111861

[64] Hong F, Wei B, Chen L. Preliminary study on biosynthesis of bacterial

nanocellulose tubes in a novel double-silicone-tube bioreactor for potential vascular prosthesis. BioMed Research International. 2015;**2015**:1-9

[65] Bao L, Tang J, Hong FF, et al. Physicochemical properties and in vitro biocompatibility of three bacterial nanocellulose conduits for blood vessel applications. Carbohydrate Polymers. 2020;**239**:116246

[66] Liu G, Ma M, Meng H, et al. In-situ self-assembly of bacterial cellulose/poly(3,4-ethylenedioxythiophene)-sulfonated nanofibers for peripheral nerve repair. Carbohydrate Polymers. 2022;**281**:119044

[67] Wei Z, Hong FF, Cao Z, et al. In situ fabrication of nerve growth factor encapsulated chitosan nanoparticles in oxidized bacterial nanocellulose for rat sciatic nerve regeneration. Biomacromolecules. 2021;**22**:4988-4999

[68] Tummala GK, Joffre T, Lopes VR, et al. Hyperelastic nanocellulose-reinforced hydrogel of high water content for ophthalmic applications. ACS Biomaterials Science & Engineering. 2016;**2**:2072-2079

[69] Wang J, Gao C, Zhang Y, et al. Preparation and in vitro characterization of BC/PVA hydrogel composite for its potential use as artificial cornea biomaterial. Materials Science and Engineering: C. 2010;**30**:214-218

[70] Yang J, Zhu Z, Liu Y, et al. Double-modified bacterial cellulose/soy protein isolate composites by laser hole forming and selective oxidation used for urethral repair. Biomacromolecules. 2022;**23**:291-302

[71] Yang J, Wang L, Zhang W, et al. Reverse reconstruction and bioprinting of bacterial cellulose-based functional total intervertebral disc for therapeutic implantation. Small. 2018;**14**:1702582

Section 3

Nanocellulose Composites and Their Use

Chapter 6

Sources and Applications of Nano Cellulose Tailored Materials

Moumita Das Kirtania, Subhasis Chakraborty, Anwesha Barik and Sanjay Dey

Abstract

Cellulose is one of the most economical, biodegradable, and biosafe components found in nature. It is extracted from biomass found in forests or crops being treated enzymatically, chemically or mechanically. The extracted cellulose on acid hydrolysis and other mechanical treatment yields bacterial cellulose, nano-fibrillated cellulose, and cellulose nanocrystals. Nanosized cellulose can be attributed to the size reduction of the polymer chains in cellulose from micro to nanoscale. The size range was found suitable from 1–100 nm to be called nanosized cellulose. Nano cellulose hogged much limelight in the modern era due to its low toxicity, biocompatibility, and biodegradability. Due to the rapid evolution in this field, it is an obvious need to synthesize nano cellulose from different sources for its huge potential in pharmaceuticals and other industries. The tiny size made the nano cellulose mechanically strong and stable thus rendering it suitable for application, especially in pharmaceuticals, drug delivery, tissue engineering, and regenerative medicine. Recent research has been focused on the development and applications of nanocellulose products due to their eco-friendly nature and diversity of its application. However, there are challenges too, related to its scale-up, cost, and stability may be registered. This chapter will further discuss in detail the synthesis and preparations of nanosized cellulose and its theragnostic applications.

Keywords: biomaterial, nano cellulose, sources of cellulose, theragonisitc applications, synthesis of nano cellulose

1. Introduction

Cellulose is one of the most widespread and abundantly available biopolymers in the biosphere, and can be easily obtained from algae, fungi, bacteria, and plants as well. Among all other similar polymers including chitosan, gelatin, starch, cellulose is considered one of the promising exhaustive renewable compounds mostly produced by photosynthesis [1]. The chemical structure of unmodified cellulose comprises a linear chain of D-glucopyranose units in the range of several hundred to thousands joined together by β-(1–4)-glycosidic linkages, whereas bacterial nano cellulose comprises nothing but a similar structure while the repeating unit is found to be disaccharide cellobiose [2]. Due to the natural hierarchy of the chemical structure in cellulose owing to its nano fibrillar components, this seemingly contributes to the unique mechanical strength, and elastic properties of the higher plant cells. Its unusually high strength-to-weight ratio also prohibits a dimensional change in response to water absorption in large amounts, shrinkage, and swelling as evident in common aqueous and terrestrial flora and fauna [3].

Consecutively reducing resources of non-renewable products is shifting the research paradigm towards the establishment of sustainable resources, which led to further purification, modification, and destructing of cellulose employing mechanical as well as chemical strategies. Chemical degradation and mechanical degradation lead to the formation of cellulose nanocrystals and cellulose microfibrils respectively. Such degradation of natural fibers separates the highly crystalline cellulose microfibrillar bundles from the remaining imperfect amorphous residue having pits and nodes [4].This homogeneous dispersion reportedly confers much superior elastic modulus and mechanical properties. However, the mechanical reinforcement can be slightly compromised for these nanocomposites since it also possesses self-associating properties due to the abundance of surface hydroxyl groups despite its ability to form load-bearing stiff structures interpenetrating suitable polymer matrix [5, 6]. The stiffness was found to increase with the aspect ratio of the nanostructures present within the microfibrils. Concerning optical properties, loading or mechanical compression of the nanocellulose fibers prevents light scattering rendering it optically transparent while the recovery of nanofibers subject to reversal of stress forms open spaces and interstices allowing light scattering [7]. Lastly, the high surface-to-volume ratio, and nanosize reduce the permeability of cellulose furthermore by enhancing the interaction of the particles with the surrounding polymer thereby reducing the mobility of chain segments, diffusivity, and reactivity to a great extent [8].

These aforementioned exclusive features of nano cellulose offer a great diversity in its application ranging from the fabrication of packaging materials to extend food shelf-life to the formulation of decontaminants, wood-adhesives, and adsorptive to name a few. The advanced medical technology has been utilizing this magical substance for the fabrication of biosensors and bio-imaging facilities [9]. The nanofibrous cellulose sheet not only offers all the advantages of conventional paper-based biosensors but this is also devoid of surface roughness, optical opaqueness, increased thermal expandability, and inability to fold. One of its several noteworthy applications includes the accurate and precise detection of *S. aureus* bacterial species for any infectious disease [10]. This article, however, gives an overview of the sources and extraction procedure of nano cellulose in its various forms along with its multidimensional usage.

2. Sources of nano cellulose

Nano cellulose can be prepared from several sources which contain cellulose. Nano cellulose is prepared from fibers of cellulose [11]. Generally, cellulose can be extracted from bacterial and algal sources, biomass from agriculture, tunicates, soft or hardwood, plants, etc. [12].

The concentration of cellulose is similar in the above-mentioned species, however, their structures may vary. For example, in plants cellulose possesses about 44–65% of crystals and has a diameter ranging from 13–22 microns [13]. However, the bacterial cellulose possesses about 90% crystals and is nano-structured with a diameter ranging from 10 to 100 nm [14]. Among the three classes of nano cellulose, nanofibrils have a length of a few micrometers and diameters ranging from 5 to 60 nm. Nanocellulose has both crystalline and amorphous forms and that can be generally extracted from plant woods by oxidizing technique [15, 16]. However, the nanocrystals are mostly crystalline with a needle shape, shorter in length ranging from 20 to 100 nm and 5 nm in diameter. Acid hydrolysis is used to extract the nanocrystals [17]. Again the nano cellulose obtained from bacterial sources has a ribbon-shaped structure, length of a few microns and diameter ranging from 20 to 100 nm [11].

2.1 Nano cellulose extracted from agriculture waste biomass

Cellulose has been extracted from agricultural wastes like pineapple peel, rice and wheat stalks, oat husk, straw, leaves, sisal, etc. using chemical methods like treatment with alkali, acid hydrolysis and bleaching. One of the most commonly used resources is sugarcane. After sugar is extracted from sugarcane the biomass left is used for extracting cellulose by chemical treatment or irradiation. The agro waste may also contains fiber, lignin, hemicelluloses, sugar, etc. apart from cellulose. Cellulose is also obtained from dietary fibers extracted from plants. Various parts of the plants can yield cellulose, for example, peel, pomace, stems, leaves, reeds, bast or roots, etc. Depending on the type of material, the extraction process differs and the use of acid, alkali, enzymes, mechanical methods, etc. Further, the structure of the cellulose obtained as a product depends not only on the type of raw material but also on the time for reaction [18].

2.2 Nano cellulose extracted from forest wood

Most commonly fibers from wood pulp are used to extract cellulose. The cellulose extracted from the wood pulp has highly cellulose content with high mechanical strength & good ductile properties. The wood pulp can be derived from both hard and softwoods.

Cellulose can be derived from the wood pulp of the trees like pine or eucalyptus, etc. in forest reserves by various chemical methods including acid treatment or oxidation and by biological methods by using fungi on softwoods [19].

2.3 Nano cellulose extracted from waste paper

A good amount of waste paper is found in the form of newspapers, newsprints, etc. from which nano cellulose has been extracted. Nanocrystals have been prepared through chemical treatment like acid hydrolysis. Nanofibers have been prepared by combining both chemical and mechanical methods like acid hydrolysis, ultrasonication or centrifugation.

2.4 Nano cellulose extracted from animal sources

Nano cellulose can also be prepared using crabs, prawns or tunicates. Nanofibers have been derived from prawns or crabs by mechanical methods. Further cellulose is also extracted from tunicates like *Halocynthiapapillosa, Metandroxarpauedai, Halocynthiaroretzi*, etc., which are marine invertebrate animals. Their outer tissue also called tunic contains crystalline cellulose in pure form [20].

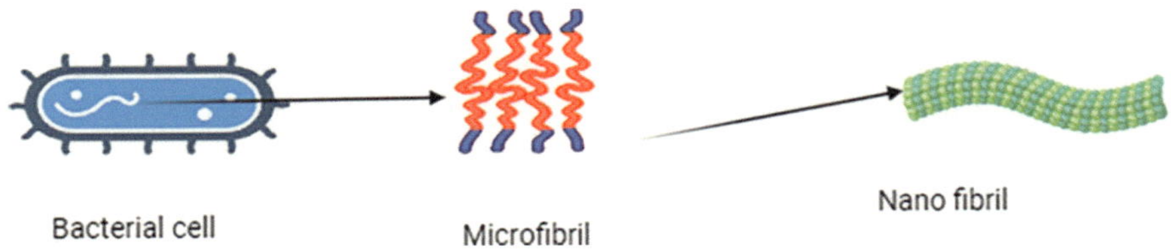

Figure 1.
Extraction of nano fibril from bacteria and algae.

2.5 Nano cellulose extracted from bacterial/algal sources

Among the algal species, cellulose is generally extracted from *Cladophora*, which is also known as green algae. The fermentation of bacterial species like *Sarcina, Agrobacterium, Gluconacetobacter*and *Rhizobium*, etc. yields nano cellulose as an extracellular product [21]. The production of bacterial cellulose was reported in both synthetic and non-synthetic media by Gluconacetobacter, Agrobacterium, and Sarcina through oxidative fermentation. The gram negative Gluconacetobacter is one of the most important one to produce bacterial cellulose in a large quantity. After starting the bacterial cellulose synthesis, the next process initiates to form the glucose chains in to the bacterial cell & it leaches out through the microscopic cellular pore and again accumulate to form the microfibrils, that leads to cellulose strip formations. That cellulose fibrils then form the web like structures & create a highly porous matrix. The cellulosic chains are attached together by the hydrogen bonding. The interesting fact regarding the bacterial cellulose is that it is 100 times smaller than the plant mediated cellulose. Many researchers & companies have developed bacterial cellulose in treating of wound, high quality excipients of papers, materials of fibre glasses etc (**Figure 1**).

3. Preparation and application of nano cellulose

Natural polymers are widely applied materials for different purposes. Due to the green approaches, they are the automatic choice of application over synthetic polymers. These green composites can be registered in different applications, especially in pharma, food, packaging, building materials and the automobile industry. Cellulose is one of them. It has comparatively a broader perception which can be found in different natural fibers such as cotton, jute, sugarcane bagasse, etc. That can be found in different bacteria, fungi and sea animals also [22–24]. Cellulose can be easily extracted on different scales such as micro-scale and nano-scale. Nano cellulose can be extracted as nano-structured cellulose fiber with a definite size range of less than 100 nm [25]. A cellulose fiber can be synthesized with the accumulation of a single cellulose fiber with a definite size range 25–30 μm in diameter. The single cellulose fiber is manufactured by the accumulation of several microfibers with a definite size range 0.1–1 μm. Nanofiber is too small with a sigh range 10–70 nm in diameter [26]. Due to its uniform tiny size and tremendous mechanical stability nano-sized cellulose is getting more interest in the present era. Different extraction procedures and other techniques are responsible to vary the characteristics of nano cellulose such as crystallinities, surface area and mechanical strength [27]. For example, the synthesis process of nano cellulose with the application of sulfuric acid (H_2SO_4) requires the selective hydrolysis of the amorphous cellulose zone as a result it forms pure crystalline particles with definite dimensions. In addition, on the surface of nano-sized cellulose particles negatively charged sulphate groups are attached through the accumulation of nanoparticles in aqueous suspensions due to the electrostatic repulsion forces. Due to the synthesis process and preparation technique of nano cellulose which hold up its dimension, composition and mechanical strengths, it can be categorized into three parts

i. Cellulose nanocrystals it is also known as cellulose whiskers.

ii. Cellulose nanofibers it is also known as cellulose nanofibers.

iii. Bacterial cellulose (BC) and electrospun cellulose nanofibers (ECNFs).

The first two phases mentioned cellulose nanocrystals and cellulose nanofibers are formed due to the fragmentation of cellulose fibers at the nano-sized level. This nano cellulose preparation technique follows the top-to-bottom approach, in the next phase such as bacterial cellulose and ECNF sugars with low molecular weight or dissolved cellulose is formulated through the bacteria or electrospinning, respectively, this process follows the opposite of the first phase which is bottom to top approach. As a result, the scale, of the process of Bacterial cellulose (BC) and electrospun cellulose nanofibers (ECNFs) is always a difficult challenge in industry [28].

4. Nano cellulose synthesis methods

4.1 Pretreatment method

Pretreatment method of plant cellulose is a widely accepted technique that decreases the required energy of mechanical nano fibrillation to accelerate the degree of nano fibrillation process. In the nano-fibrillation technique, the major complaint is regarding the huge energy consumption to produce the nanofibril. As a result, pretreatment is an important step to overcome this drawback. The pretreatment process is also a major step to producing more nanofibers through the nano-fibrillation technique. It is very common that two major issues are specifically noticed in pretreatment method specially during the mechanical fibrillation of cellulose i. fibril aggregation in to the disintegration device when the slurry was pumped & another is high energy consumption associated with the fiber de lamination. High energy sources & input is important to overcome the barriers such as e interfibrillar hydrogen bonding. Different scientifical works also proved that the potent or actual pretreatment method actually reduces the energy at least 20–30 folds. The choice of the pretreatment method is actually depending upon the sources of the cellulosic fibers & its detailed morphology. For instance, the pretreatment method is really effective to remove the non-cellulosic materials from the plant such as (hemicellulose, lignin, etc). Pretreatment of tunicate associates the execution of the protein matrix, isolation of the mantel, and the isolation of individual cellulose fibrils. Pretreatment of algae basically involves the separation of the matrix substance from the cell wall of algae, whereas in pretreatment method of bacterial nitrocellulose expresses the procedure of separation of bacteria & other impurities from the prepared slurry. This pretreatment procedure is very vital one especially to produce the nano cellulose because it can alter the structural orientation, crystallinity & polymorphism [29].

4.2 Enzyme hydrolysis

Some enzymes are able for selective hydrolysis for example laccase can modify the lignin content or hemicellulose content without any alteration of the main cellulose content. Cellulose fibers are made with different compounds with sound structures as a result, a particular enzyme cannot alter the structure of a nanofiber. There are some of the enzymes are required to decompose the extra cellulose compounds [3].

1. Cellobiohydrolases: It strikes mostly on crystalline cellulose.

2. Endoglucanases: These strike mostly the affected structure of the cellulose.

A group of scientists Pääkkö and Ankerfors has shown the pathway to prepare nano cellulose from the bleached softwood pulp [30]. In this preparation method the enzymatic hydrolysis technique was used to prepare nano cellulose and after that for better nano structure homogenization and purification process followed. Another finding from this analytical procedure is mild hydrolysis with the application of endoglucanase has escalated the aspect ratio of nano cellulose without any conflict as compared to the acid hydrolysis technique. Another advantage of this pretreatment method is that it elevates the solid level thereby it permits a smooth path during HPH processing [31]. Other scientists also worked with this technique and found the results of TEM to be very suitable. 90% of the prepared nanofibers have given a diameter of less than 50 nm. It also gives a proper and better aspect ratio as compared to the untreated version [32].

4.3 Alkaline–acid

Before the mechanical isolation technique of nano cellulose, alkaline acid pretreatment is playing a major role to solubilize lignin, hemicellulose, and pectin. This method is following different steps which are given below.

a. It is mandatory to soak the fibers with a concentration of 12–17.5% sodium hydroxide solution at least for 2 hours. That can increase the surface morphology of the nanofibers and make the hydrolysis process easier.

b. Soaking of the fiber in 1 M hydrochloric acid solution at a temperature of 60–75°C. This technique can solubilize the hemicellulose most efficiently.

c. For breaking the linkage between carbohydrate and lignin structure sodium hydroxide 2% solution is treated at 60–80°C.

Alkaline pre-treatment is an effective technique that can increase the yield value of different nano cellulose from 42–80% [33].

4.4 Ionic liquids

Thermal stability, chemical stability & low vapor pressure are adopted as some of the special facility in the ionic liquid process. This is an exciting technique for formulation scientists to use as a solvent for dissolving cellulosic materials. Scientists reported some of the work where 1-butyl-3-methylimidazolium chloride [(Bmim)Cl] as an ionic liquid (solvent) to treat sugarcane bagasse. After that, the homogenization process was followed to prepare nano cellulose [34].

4.5 Mechanical process

Mechanical treatment is an important step for the defibrillation of cellulosic materials. The pretreatment process is the primary process either by chemical or enzyme included to make it easier. The application of chemical treatment has done for widening the surface space between the hydroxyl group, which elongates the internal surface, it also alters the crystallinity and create fragment of cellulose hydrogen bonds. Thereby, the surface area is increased which boots up the reactivity of the nanofiber [35]. Plenty of methods are available to prepare the nano cellulose from cellulose fiber, such as homogenization, grinding, microfludized technique, cyrocrushing and high-density ultrasonication technique [36].

4.6 Homogenization technique

This process was introduced in 1983, it is one of the simplesttechnique to process the nano fibers from cellulose fibers. This process does not require any organic solvent [37]. In homogenization technique the cellulose pump is required & it is designed with a very tiny nozzle withremarkable high pressure. Mostly the high impact and shear forces are applied for preparing the nano cellulose. The forces generate high shear rates as a result the nano scale fibers are formed [38]. Some good numbers of peer-reviewed journals reported that bleached sugar beet and the extraction process of prickly has been done through this process. But still there is some limitation notified in this process one of them is clogging. To bypass this clogging issue the cellulosic fiber has to be chopped into very small pieces before introducing in homogenizer [39].

4.7 Micro fluidization technique

The working procedure of microfludization technique was more or less the same as the homogenization technique. Microfluidizers have been applied in nano cellulose preparation to enhance the pressure as an intensifier pump. Both forces worked into the interaction chamber (Impact and shear forces) against colliding streams to de-fibrillate the cellulose fibers. Higher numbers of hydroxyl groups (-OH) were observed in cellulose fibrils due to the higher surface area. The size of the nano cellulose was determined by the actual number of chopped fibers through the homogenizer [40].

4.8 Grinding

This technique is another technique to prepare nano-size cellulose from normal cellulose fibres. This process works with the impact and attrition principle. In this context, the pulp of the cellulose was introduced between two stones. One stone is static and another stone is movable. The grinding mechanism is applied to create the shear forces which break the hydrogen bond & damage the cell wall completely. As a result the nano sized cellulose is formed from the normal cellulose. Wang et al. reported through their scientific work after the application of a stone grinder the normal pulp of bleached eucalyptus can be modified to nano cellulose [41]. The attrition and friction process from the stone grinder generates heat which can evaporate water and creates more solid contents. It requires a long time (11–12 h) to create actual fibrillation energy to make nano-scale products [42].

4.9 Cyrocrushing

This method is another method that can create the fragmentation into the cell wall of the normal cellulose to prepare nano-sized cellulose. The cellulose is introduced into the water to absorb the moisture into the cellulose cavity. The soaked cellulose is deeped into the liquid nitrogen carefully that makes solid content of water, and after that, it is crushed into the mortar pastle [43]. Due to the application of the high impact force (impact) on the pre-freeze cellulosic fibers that break into fragments and as a result, create nano cellulose. Scientists have found 50–100 nm nanofibers from the soya stock after using this technology along with high-pressure homogenization technology [44].

4.10 High-intensity ultrasonication technique

Ultrasonication technique is one of the most common techniques to prepare different nanoparticles along with various conditions. This process includes hydrodynamic power with the effect of ultrasound as a result it can create nanofibers. Different studies have prepared nano cellulose fibers through the high-intensity ultrasonication technique with the effect of temperature, pressure, size, time, and distance from the probe tip on the degree of fibrillation [45].

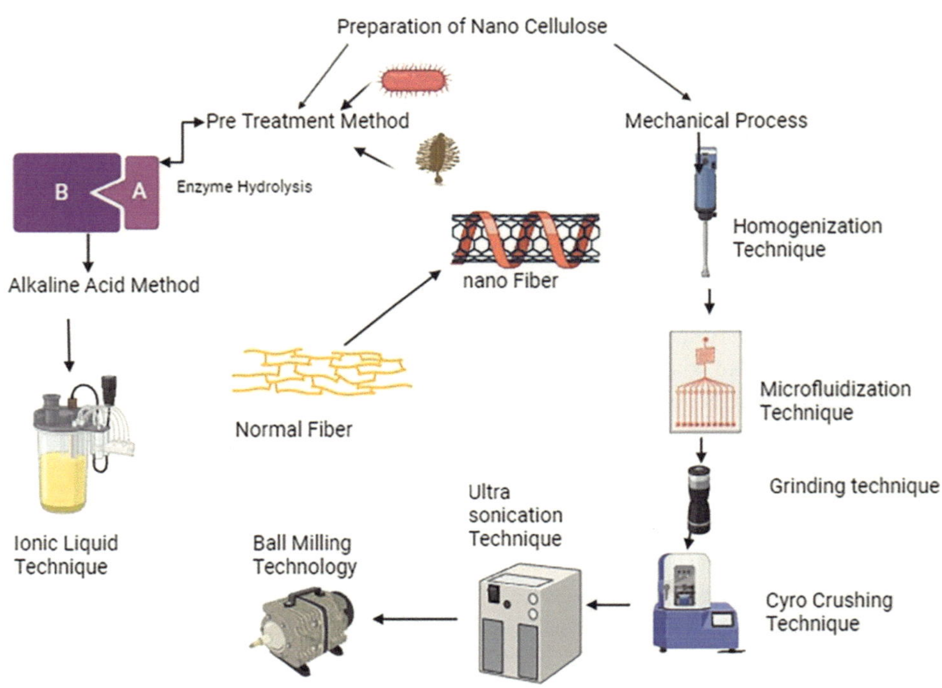

Figure 2.
Schematic diagram of preparation of nano cellulose.

4.11 Ball milling

It is one of the oldest methods to prepare nano cellulose or nanofiber through the application of impact and attrition technique. The hollow chamber of the ball mill contains balls specially made for ceramics or stainless steel. The cylinder rotates with the help of the motor so that a high-energy collision occurs with the ball and cellulose fibers. As a result, the size reduction was taken place and the normal cellulose fiber turns up to nano cellulose (**Figure 2**) [46].

5. Application of nano cellulose in industry

Nano cellulose is one of the materials which is supposed to be the alternative to synthetic material and is eco-friendly. It is a comparatively new biomaterial that has different roles in industries such as the paper industry, packaging industry, automotive, electronic, medical, and wastewater treatment, etc. Due to its excellent mechanical and chemical stability, it is one of the choices in the industry. Some interesting equipment such as windmills is made with nano cellulose due to its stable mechanical structures and huge strength. In the paper industry nano cellulose is used because of its soft nature, folding capacity and clarity. Due to its less toxicity and mechanical strength, it is also applied in the medical industry, especially for operation theater. Scientists have found that nano cellulose is applied in dressing and it has unique biocompatibility in skin drafting and tissue engineering rather than the others [47]. Nano cellulose dressing can be well-designed and safe and easily detachable after skin recovery. However, nano cellulose is presently a positive choice for different industries such as automotive, pharmaceuticals, paper & ink & cosmetics etc. It can also be applied as a membrane of high accuracy ultrafiltration technique, ion exchange & fuel cell etc. Nano cellulose is also one of the major choices for bio nano composites especially to prepare different film for transdermal application, absorption & preparation of medicated foam. Researchers reported that scaffolding of polymeric potato starch matrix & nano fiber as a rein forcing agent can improve the mechanical properties of nano composites remarkably. At present day it is established that nano cellulose is one of the best choices as a reinforcing agent. Another delivery of the nano cellulose is also notified by some peer-reviewed journals such as drug delivery, implantation, tissue drafting, tissue engineering, etc.

5.1 Application of nano cellulose in paper industry

Application of nano cellulose in paper industry was deputed to replace the application of the hazards petrochemicals. More than 100 million tons of commercially available cellulose fiber was introduced in the paper industry every year. There are various steps also available to produce fine paper such as arrangement of paper components, wet refining, preparation of fine sheet, drying, calendaring, Finishing and polishing. A mandatory step for paper making is to refine the cellulose fiber in a moist medium is common and best to make a quality paper. Nano cellulose can increase the tensile strength of the paper when applied. These fine paper sheets have better mechanical properties than normal paper. But one major drawback of this process is too costly [48]. Nanocellulose is blessed with different properties one of them is tiny size with nano scale & gas & water barrier properties make it stronger applicant for paper industry. Due to these properties, it can prepare a definite thin film with good mechanical strength, which is one of the requirements for paper industry. Moreover, it is extracted from the natural sources and it is nontoxic& non irritative in nature as a result it is also applied in food packing especially for export purposes. It is reported in some peer reviewed scientific literature that application of nano clay & poly lactic acid having nano cellulose one of the rein forcing agent has definite mechanical strength to produce a proper texture for food packaging as a barrier of oxygen & water specially.

5.2 Application of nano cellulose in medical industry

In biomedical industry application of nano cellulose is comparatively new & in modern era it has got a huge appreciation in medical as well as pharma industry. It is observed through different research that electrospunpolylactide-co-glycolide (PLGA) is more effective than normal PLGA for skin tissue engineering. Electro-spun polylactide-polyglycolide has better cell compatibility in skin tissue engineering this affects the adhesion, spreading and proliferation of the fibroblasts. The mechanical stability of different engineered tissues also increased due to the application of nano cellulose. In vascular tissue engineering replacement and repairing of tissues are very common. Lacking endothelial cell linings, it causes thrombosis when reconstruction of tiny diameter blood vessels. As a result, endothelialization is an essential parameter to produce artificial blood vessels. Among some natural molecules collagen, hyaluronic acid has tremendous potential and good biocompatibility but they have lacuna in mechanical strength [49]. Among some synthetic polymer like polyethylene terephthalate is mechanically stable but have an issue with the affinity towards the cell and tissue that causes severe complication in scaffolding. Nano cellulose is chosen because of its tiny size and high biocompatibility with others and also biodegradable. In neural tissue engineering also nano cellulose has tremendous potency. Its mechanical stability, porosity and sound biocompatibility make it advantageous over the others (**Figure 3**) [50].

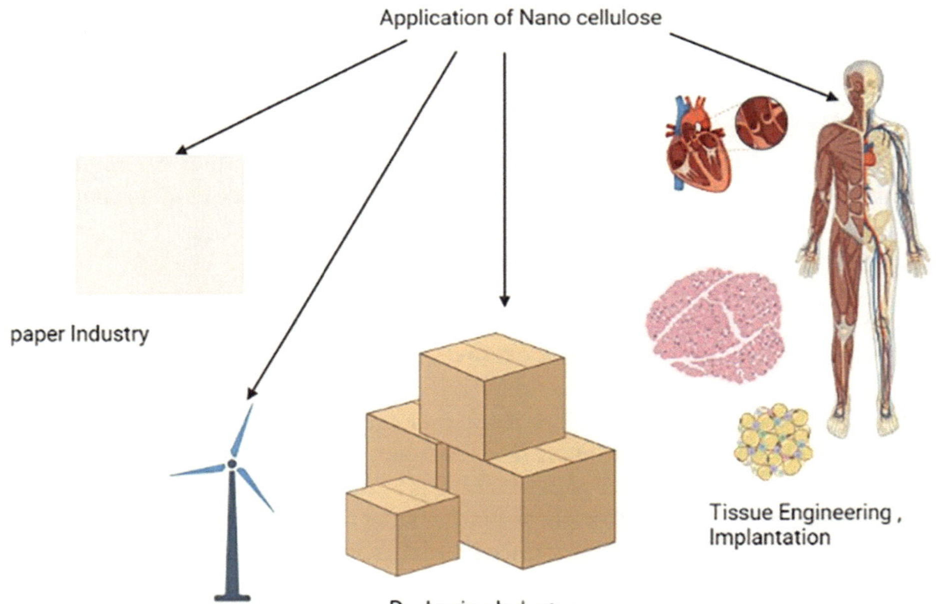

Figure 3.
Application of nano cellulose in different industry.

6. Future prospects of nano cellulose

Several research studies and innovations have been continuing with nano cellulose and its modified forms for the last 20 years. Several limitations have been overcome and modifications are done as per the requirement of the industry with eco-friendly features. In this present book chapter, nano cellulose has been explored concerning its true potentiality towards the green concept as well as full its ecofriendly nature. It has some tremendous properties which make nano cellulose always better suited compared to others such as high surface area, better chemical and mechanical stability, proper anisotropic shape, etc. These properties make nano cellulose an excellent addition in biomedical and other fields with high potentiality and more durability. Due to the cost-effective and eco-friendly behavior, it can get a lodge of new applications as per the industry requirement. It is also observed that the applications of nano cellulose in different industries are in global demand. As a result, the interest in nano cellulose preparation and modification is an automatic choice in the future. Therefore, it is an obvious need to facilitate and improve the technology to prepare and modify nano cellulose from lab scale to industrial scale for future application. This technology should be improved due to the transition gap between lab scale to industry and finally to the market. The future target of nano cellulose research is to fill the gap between the transition state of technology transfer and scale-up. Optimization of the whole scale-up process and its stability should be justified in future studies. The life cycle assessment of nano cellulose-based products should be justified for future betterment and a futuristic approach. Innovation and technology should look after energy consumption and time management to produce nano cellulose-based products from cellulose and its derivatives. Despite all the limitations, cost and other amenities we hope that the formulation scientists will adopt the newer technologies to prepare nano cellulose-based materials to improve people's quality of life in the near future [51].

7. Conclusion

In the whole chapter, we have discussed the sources, preparation methods and applications of nano cellulose. We have observed that there are several sources from which cellulose fibers can be collected and further nano cellulose may be produced by various means. The various types of nano cellulose can be developed from various sources and methods whether physical, chemical or biological. Nano cellulose has highly beneficial properties concerning durability, biodegradability, and mechanical strength and thus finds an end number of applications in pharmaceuticals, biopharmaceuticals, paper or food industry. Keeping in mind the huge number of applications and the growing interest of researchers to work with nano cellulose, proper development and scale-up methods need to be established so that nano cellulose may be applied cost-effectively and productively in products adding value to human life.

Author details

Moumita Das Kirtania[1], Subhasis Chakraborty[2], Anwesha Barik[3] and Sanjay Dey[2]*

1 Techno India University, Bidhannagar, Kolkata, West Bengal, India

2 DmbH Institute of Medical Science, Hooghly, West Bengal, India

3 Department of Pharmaceutical Technology, School of Medical Sciences, Adamas University, Barasat, West Bengal, India

*Address all correspondence to: sanju1980dey@gmail.com

IntechOpen

© 2023 The Author(s). Licensee IntechOpen. This chapter is distributed under the terms of the Creative Commons Attribution License (http://creativecommons.org/licenses/by/3.0), which permits unrestricted use, distribution, and reproduction in any medium, provided the original work is properly cited.

References

[1] Sundarraj AA, Ranganathan TV. A review on cellulose and its utilization from agro-industrial waste. Drug Invention Today. 2018;**10**:89-94

[2] McNamara JT, Morgan JL, Zimmer J. A molecular description of cellulose biosynthesis. Annual Review of Biochemistry. 2015;**84**:895-921

[3] Dufresne A. Nano cellulose: A new ageless bionanomaterial. Materials Today. 2013;**16**:220-227

[4] Dufresne A. Nano Cellulose: From Nature to High Performance Tailored Materials. Walter de Gruyter GmbH and Co KG; 2017

[5] Bras J, Viet D, Bruzzese C, Dufresne A. Correlation between stiffness of sheets prepared from cellulose whiskers and nanoparticles dimensions. Carbohydrate Polymers. 2011;**84**:211-215

[6] Iwamoto S, Nakagaito AN, Yano H. Nano-fibrillation of pulp fibers for the processing of transparent nanocomposites. Applied Physics A: Materials Science and Processing. 2007;**89**:461-466

[7] Nogi M, Iwamoto S, Nakagaito AN, Yano H. Optically transparent nanofiber paper. Advanced Materials. 2009;**21**:1595-1598

[8] Lavoine N, Desloges I, Dufresne A, Bras J. Microfibrillated cellulose—Its barrier properties and applications in cellulosic materials: A review. Carbohydrate Polymers. 2012;**90**:735-764

[9] Golmohammadi H, Morales-Narváez E, Naghdi T, Merkoçi A. Nano cellulose in Sensing and Biosensing. Chemistry of Materials. 2017;**29**:5426-5446

[10] Langari MM, Nikzad M, Labidi J. Nano cellulose-based sensors in medical/clinical applications: The state-of-the-art review. Carbohydrate Polymers. 2023;**304**:120509

[11] Kaur P, Sharma N, Munagala M, Rajkhowa R, Aallardyce B, Shastri Y, et al. Nano cellulose: Resources, physio-chemical properties, current uses and future applications. Frontiers in Nanotechnology. 2021;**3**

[12] Jonoobi M, Oladi R, Davoudpour Y, Oksman K, Dufresne A, Hamzeh Y, et al. Different preparation methods and properties of nanostructured cellulose from various natural resources and residues: A review. Cellulose. 2015;**22**:935-969

[13] Ververis C, Georghiou K, Christodoulakis N, Santas P, Santas R. Fiber dimensions, lignin and cellulose content of various plant materials and their suitability for paper production. Industrial Crops and Products. 2004;**19**:245-254

[14] Gorgieva S, Trček J. Bacterial cellulose: Production, Modification and Perspectives in Biomedical Applications. Nanomaterials. 2019;**9**

[15] Habibi Y, Lucia LA, Rojas OJ. Cellulose nanocrystals: Chemistry, self-assembly, and applications. Chemical Reviews. 2010;**110**:3479-3500

[16] Jonasson S, Bünder A, Niittylä T, Oksman K. Isolation and characterization of cellulose nanofibers from aspen wood using derivatizing and non-derivatizing pretreatments. Cellulose. 2020;**27**:185-203

[17] Phanthong P, Reubroycharoen P, Hao X, Xu G, Abudula A, Guan G. Nano

cellulose: Extraction and application. Carbon Resources Conversion. 2018;**1**:32-43

[18] Wang J, Han X, Zhang C, Liu K, Duan G. Source of nano cellulose and its application in nanocomposite packaging material: A review. Nanomaterials. 2022;**12**:3158

[19] Noor SM, Anuar AN, Tamunaidu P, Goto M, Shameli K, Halim MHA. Nano cellulose production from natural and recyclable sources: A review. IOP Conference Series: Earth and Environmental Science. 2020;**479**:012027

[20] Shak KPY, Pang YL, Mah SK. Nano cellulose: Recent advances and its prospects in environmental remediation. Beilstein Journal of Nanotechnology. 2018;**9**:2479-2498

[21] Gupta GK, Shukla P. Lignocellulosic biomass for the synthesis of nano cellulose and its eco-friendly advanced applications. Frontiers in Chemistry. 2020;**8**

[22] Asim M, Abdan K, Jawaid M, Nasir M, Dashtizadeh Z, Ishak MR, et al. A review on pineapple leaves fibre and its composites. International Journal of Polymer Science. 2015;**2015**:950567

[23] Isogai A. Wood nano celluloses: fundamentals and applications as new bio-based nanomaterials. Journal of Wood Science. 2013;**59**:449-459

[24] Oun A, Rhim J-W. Characterization of nano celluloses isolated from Ushar (Calotropis procera) seed fiber: Effect of isolation method. Materials Letters. 2016;**168**

[25] Osong SH, Norgren S, Engstrand P. Processing of wood-based microfibrillated cellulose and nanofibrillated cellulose, and applications relating to papermaking: a review. Cellulose. 2016;**23**

[26] Siqueira G, Bras J, Dufresne A. Cellulosic bionanocomposites: A review of preparation, properties and applications. Polymers. 2010;**2**:728-765

[27] Moon RJ, Martini A, Nairn J, Simonsen J, Youngblood J. Cellulose nanomaterials review: structure, properties and nanocomposites. Chemical Society Reviews. 2011;**40**:3941-3994

[28] Lin N, Dufresne A. Nano cellulose in biomedicine: Current status and future prospect. European Polymer Journal. 2014;**59**:302-325

[29] Gilbert HJ, Knox JP, Boraston AB. Advances in understanding the molecular basis of plant cell wall polysaccharide recognition by carbohydrate-binding modules. Current opinion in structural biology. 2013;**23**:669-677

[30] Boerjan W, Ralph J, Baucher M. Lignin biosynthesis. Annual review of plant biology. 2003;**54**:519-546

[31] Saba N, Paridah TM, Abdan K, Ibrahim N. Preparation and characterization of fire retardant nano-filler from oil palm empty fruit bunch fibers. BioResources. 2015;**10**

[32] Spence K, Habibi Y, Dufresne A. Nano cellulose-Based Composites. Springer; 2011. p. 179-213

[33] Henriksson M, Berglund LA, Isaksson P, Lindström T, Nishino T. Cellulose nanopaper structures of high toughness. Biomacromolecules. 2008;**9**:1579-1585

[34] Ioelovich M, Leykin A. Nano-cellulose and its application. SITA. 2004;**6**:17-25

[35] Siqueira G, Tapin-Lingua S, Bras J, Da Silva PD, Dufresne A. Morphological investigation of nanoparticles obtained from combined mechanical shearing, and enzymatic and acid hydrolysis of sisal fibers. Cellulose. 2010;**17**:1147-1158

[36] Roman M, Winter WT. Effect of sulfate groups from sulfuric acid hydrolysis on the thermal degradation behavior of bacterial cellulose. Biomacromolecules. 2004;**5**:1671-1677

[37] Liu D, Zhong T, Chang PR, Li K, Wu Q. Starch composites reinforced by bamboo cellulosic crystals. Bioresource technology. 2010;**101**:2529-2536

[38] Filpponen I, Argyropoulos DS. Regular linking of cellulose nanocrystals via click chemistry: Synthesis and formation of cellulose nanoplatelet gels. Biomacromolecules. 2010;**11**:1060-1066

[39] Zhang L, Tsuzuki T, Wang X. Preparation of cellulose nanofiber from softwood pulp by ball milling. Cellulose. 2015;**22**:1729-1741

[40] Chen P, Yu H, Liu Y, Chen W, Wang X, Ouyang M. Concentration effects on the isolation and dynamic rheological behavior of cellulose nanofibers via ultrasonic processing. Cellulose. 2012;**20**

[41] Frone A, Panaitescu D, Donescu D. Some aspects concerning the isolation of cellulose micro- and nano-fibers. UPB Scientific Bulletin, Series B: Chemistry and Materials Science. 2011;**73**

[42] Iwamoto S, Kai W, Isogai A, Iwata T. Elastic modulus of single cellulose microfibrils from tunicate measured by atomic force microscopy. Biomacromolecules. 2009;**10**:2571-2576

[43] Wang T, Drzal L. Cellulose-nanofiber-reinforced poly(lactic acid) composites prepared by a water-based approach. ACS applied materials and interfaces. 2012;**4**:5079-5085

[44] Lee S-Y, Chun S-J, Kang I-A, Park J-Y. Preparation of cellulose nanofibrils by high-pressure homogenizer and cellulose-based composite films. Journal of Industrial and Engineering Chemistry. 2009;**15**:50-55

[45] Habibi Y, Mostafa M, Vignon M. Microfibrillated cellulose from the peel of prickly pear fruits. Food chemistry. 2009;**115**:423-429

[46] Keerati URM, Corredig M. Effect of dynamic high pressure homogenization on the aggregation state of soy protein. Journal of Agricultural and Food Chemistry. 2009;**57**:3556-3562

[47] Herrick FW, Casebier RL, Hamilton JK, Sandberg KR. Microfibrillated cellulose: morphology and accessibility. In: Journal of applied polymer science. Applied polymer symposium, United States. Vol. 37. Shelton, WA United States: ITT Rayonier Inc.; Jan 1983

[48] Johnson RK, Zink-Sharp A, Renneckar S, Glasser WG. A new bio-based nanocomposite: fibrillated TEMPO-oxidized celluloses in hydroxypropylcellulose matrix. Cellulose. 2009;**16**:227-238

[49] Chakraborty A, Sain M, Kortschot M. Cellulose microfibrils: A novel method of preparation using high shear refining and cryocrushing. Holzforschung. 2005;**59**:102-107

[50] Panthapulakkal S, Sain M. Preparation and characterization of cellulose nanofibril films from

wood fibre and their thermoplastic polycarbonate composites. International Journal of Polymer Science. 2012;**2012**:381342

[51] Abdul Khalil H, Chong E, Owolabi F, Asniza M, Tye Y, Rizal S, et al. Enhancement of basic properties of polysaccharide-based composites with organic and inorganic fillers: a review. Journal of Applied Polymer Science. 2019;**136**:47251. DOI: 10.1002/app.47251

Chapter 7

Surface and Reducing End Modification of Nanocellulose to Tailor Miscibility and Mechanical Performance of Reinforced Elastomer Sustainable Composites

Jinlong Zhang, Qinglin Wu and Weiguo Li

Abstract

The development of sustainable nanocelluloses (CNCs) reinforced elastomer composites with high performance is of importance to address the plastic contamination issues, especially the ocean plastic pollution. However, achieving homogeneous dispersion and nanoscale reinforcement of CNCs in the hydrophobic elastomer matrix has been a primary challenging issue attributed to the hydrophilic feature of CNCs and agglomeration during processing. To tailor the mechanical performance of CNCs reinforced elastomer nanocomposites, surface modification of nanocellulose as the traditional way and its reducing end modification as an emerging novel method are ideal strategy to improve its overall performance. In this chapter, nanocellulose surface and reducing end modification to design high performance sustainable CNCs reinforced elastomer composites will be summarized.

Keywords: nanocellulose, sustainable composites, mechanical property, reducing end modification, surface modification

1. Introduction

Polymers commonly named as plastics have become so widely used as a consequence of safeness and low cost, and their production by 2050 is estimated to reach above 500 million tonnes. This ever- increasing production, combined with our desirability for disposability and poor mechanisms for recycling, has led to a plague of plastic in our environment as shown in **Figure 1**. Thus, plastic pollution is a big challenging issue in the world currently, especially ocean micro-plastics [2].

The research in terms of renewable resources derived sustainable composite materials is of importance to address the plastic pollution issues, especially the microplastics in the ocean [1, 3]. The utilization of renewable resources (e.g., natural rubbers and wood fibers) for the design of sustainable composites also has been attracted attentions due to the depletion of fossil resources [4]. According to the literature, the

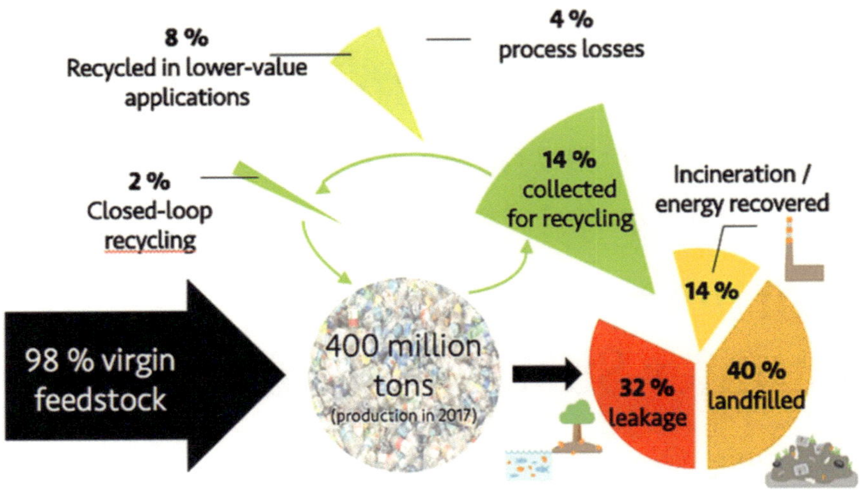

Figure 1.
Waste plastic and recycling published in 2020 [1].

unique traits of renewable resources based nanocellulose are modulus over 150 GPa and strength over 10 GPa, which is a widely used nanomaterial to design CNC high-performance composites and smart functional materials [5]. Polydimethylsiloxane (PDMS), polyurethane (PU) and polyacrylate as the commonly used elastomer materials have been widely used for surface coating, sealant, and medical cable materials. However, their low elastomer modulus is the primary drawback, which restrict their wide use in the industry fields. Therefore, CNCs are considered as an ideal renewable resource nanomaterial to tailor sustainable elastomer composite performance. These modified elastomer composites are potential for applications in biomedical devices, automotive parts, and stretchable electronics. Several elastomeric sustainable nanocomposite materials reinforced with CNCs were investigated for the last 10 years, such as PU/CNC composites [6], and these CNCs filled elastomer composites resulted in increased thermal and elastic modulus property. Despite of largely enhanced Young's modulus in the resulted CNC-reinforced nanocomposite materials, the tensile strength values were obviously decreased. The immiscibility in the CNC-reinforced elastomer sustainable composite system were the possible reasons.

Thus, to achieve the satisfied enhancement in term of mechanical property is hard attributed to the insufficient interactions at the molecular scale among CNCs and elastomer matrices [7]. The high polarity feature of CNCs resulted in its aggregation easily in the elastomeric matrices, thereby leading to the unsatisfactory compatibility [8]. Thus, it is still a challenging issue to enhance the miscibility and mechanical performance of CNC-reinforced elastomeric composite system. The use of a suitable compatibilizer is a direct way to tailor the compatibilization of CNC-reinforced elastomeric sustainable nanocomposites to establish interfacial interactions (e.g., electrostatic attractions and entanglement of molecular chains) [9]. In terms of basic principle, elastomer matrices and CNCs are compatibilized with the graft copolymer (A-g-B) or chain segments of block copolymer named as A-b-B as shown in **Figure 2**. For instance, the effect interfacial miscibility and dispersion of cellulose in the hydrophobic polyethylene (PE) matrix was achieved attributed to the chain segments in the copolymer compatibilizer interacted with both PE matrix and cellulose dispersion fillers [10]. For grafting or block copolymer noted as A-g-C/A-b-C working on the miscibility of

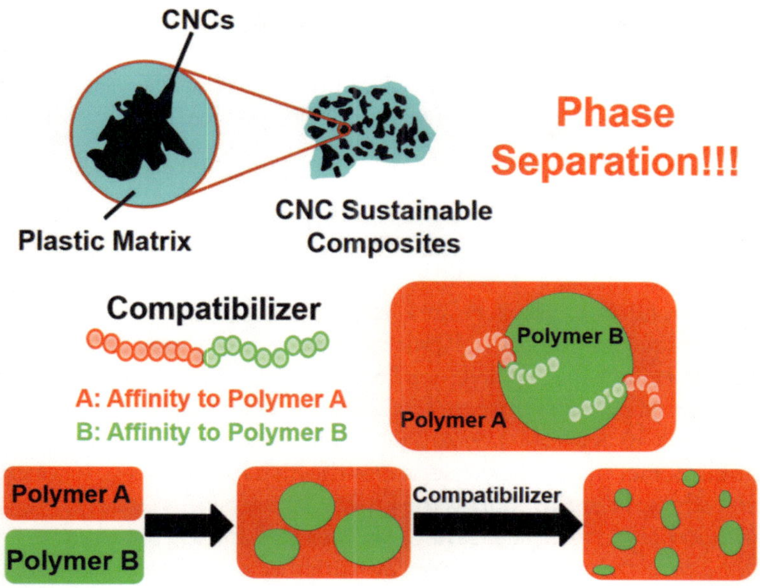

Figure 2.
Compatibilization strategy of CNCs reinforced elastomer composites.

CNC-reinforced elastomer composite system as another strategy, the basic principle is that one segment of compatibilizer miscible with CNCs, and another segment is identical with elastomer matrix. For instance, poly(L-lactic acid) (PLA)/cellulose composite system was compatibilized with PLA grafted xyloglucan [11]. Despite of the improved interfacial miscibility in CNC reinforced elastomer sustainable composite system via the compatibilizer approach, its mechanical performance is still not satisfied with industry requirements and has a large space for further improvement. Therefore, two primary strategies as the primary theme in this chapter are summarized to further tailor the interfacial miscibility and mechanical performance of sustainable CNC-reinforced elastomer composites, namely, CNC reducing end and surface modifications.

2. Surface modification of CNCs for tailoring miscibility and mechanical performance of its sustainable elastomer nanocomposites

2.1 Surface modification of CNCs

CNC surface modification strategies are primarily divided into two categories. The first approach is to construct the effective linkage among functional compounds and CNCs through the covalent or non-covalent bonds (hydrogen bond and electrostatic interaction). The second approach is to transform CNC surface properties (e.g., hydrophilicity) with the purpose of enhancing its miscibility with the elastomeric matrices. Polymer grafting (grafting from and grafting to) [12], TEMPO-mediated oxidation [13] and adsorption of surfactants [14] as well as esterification [15] as the common strategies have been systemically explored to tailor the surface property of CNCs. However, a relatively weak durability in term of a modifier absorpted onto the surface of CNCs is a big problem resulting in the surface modifications in low efficiency. The grafting copolymerization via the covalent bond linkages was employed for addressing these drawbacks. According to the literature, it is worth noting that surface grafting polymer chains, surface group derivatization and surface bonding with functional molecules are the primary theme in terms of surface modification of CNCs for the last 20 years. Please see the representative review articles in terms of CNC surface modifications [16–18].

2.2 Surface modification of CNCs on miscibility and mechanical performance of its reinforced elastomer nanocomposites

Taking CNC-reinforced polyacrylate elastomer composites as a typical example to track the technique progress of CNC surface modification for tailoring miscibility and mechanical performance of CNC-reinforced elastomer sustainable composites. The CNC-reinforced polyacrylate copolymer latex in terms of varying CNC contents was reported about 20 years ago [19, 20], while the obvious aggregation of CNCs in the polyacrylate elastomer matrix was observed as the high loadings of CNCs were added. Subsequent study was done to address the interfacial miscibility issues for the bacterial cellulose and polyacrylate latex. However, the mechanical property was obviously decreased despite of the enhancement of compatibility in the bacterial cellulose-reinforced polyacrylate latex composite system [21]. In order to address the compatibilization of CNCs and polyacrylate elastomer materials, the novel approaches named as in situ polymerization or surface grafting modification were explored [22]. The micro-phase separation and mechanical performance of CNC-reinforced polyacrylate elastomer composites via the surface grafting modification avenue were further investigated as well. For instance, to reduce the chance of hydrophilic CNC aggregation in the elastomer matrix named as polyacrylate copolymers, poly(methyl methacrylate) were chemically grafted onto the surface of rod-like CNC surface to impart its hydrophobic character [23]. The obviously enhanced mechanical performance of resulted CNCs/polyacrylate nanocomposite system were resulted from the micro-phase separation structure and good interfacial miscibility [24]. Thus, it is a feasible method to tailor the mechanical property and compatibilization of CNCs and elastomer matrix via the surface grafting modification approach. However, this conventional surface modification strategy needs to consume CNC surface hydroxyl groups, resulting in the disruption of percolation network, which was constructed via the interactions of hydrogen bonding formed from CNC surface hydroxyl groups. Thus, how to homogeneously disperse CNCs and construct strong interfacial miscibility while maintaining percolating networks of CNCs is a challenging issue. A novel approach named CNC end modification working on the aldehyde groups at the end of CNCs has been attracted attentions a few years ago. The CNC end modification as a topochemical and local modification strategy is obviously different from previous surface grafting modification. The CNC end modification is still in its infancy and demands further study compared to the hydroxyl group surface modification strategies already reported in the literature.

3. Improved miscibility and mechanical performance of sustainable CNC elastomer nanocomposites via its reducing end modification

3.1 Reducing end modification of CNCs

In comparison with hydroxyl group modification on CNC surface, it is a new perspective for cellulose chain end modification named as a regioselective reaction on multiple areas of nanocellulose particle. Thus, the active aldehyde groups on CNC end provides an avenue for the modification of CNCs. Surface hydroxyl group modification approach is obviously different from aldehyde group modification in terms of reactivity or quantity of surface hydroxyl and end aldehyde active groups. The hydroxyl functionality on the backbone of CNCs with varied reactivities (C3-OH/C2-OH < C6-OH) are primarily used for the surface modification, while aldehyde groups with high reactivity are mainly provided for the reducing end modification, but steric hindrance influence as an important factor should be considered [25]. The content of aldehyde groups on CNC end is much less than that of surface hydroxyl groups considering the available group contents for the CNC chemical modification. However, some unique functions or features enable to be introduced on CNC end through tailoring the aldehyde groups directly on one end of CNCs [26].

The CNC reducing end modifications are mainly divided into three categories according to its reaction mechanism [28]. The direct and common strategy is transforming aldehyde groups (−CHO) to carboxyl groups (−COOH) followed by the carboxyamine condensation with amino compounds by covalent bonding. For instance, the biotin functionality was selectively introduced at CNC end based on carbodiimide reaction, which includes the carbonyl group oxidation followed by the nucleophilic reaction with biotin compounds containing amine groups as shown in **Figure 3** [27].

Another feasible strategy is using aldimine condensation reaction between aldehyde group at the CNC reducing end and amino compounds. For instance, a two-step chemical reaction, namely the hydrazone reaction and subsequent amidation reaction, has been studied to selectively introduce alkyl long chains with 18-carbon on the end of CNCs as shown in **Figure 4**. The resulted oil/water Pickering emulsions with the alkyl chains (18 chains) grafted CNCs as the stabilizers showed a high emulsifying performance, and the completely cleaved 18-carbon alkyl chains enabled to be achieved with a proper acid condition, so the resulting Pickering emulsions had a pH responsive property [29]. The Knoevenagel condensation between diketone and aldehyde on the reducing end as another strategy has been studied as well. For example, the of Knoevenagel condensation has been employed to chemically decorate the acetyl groups on CNC reducing end by its reaction with beta-diketone, which

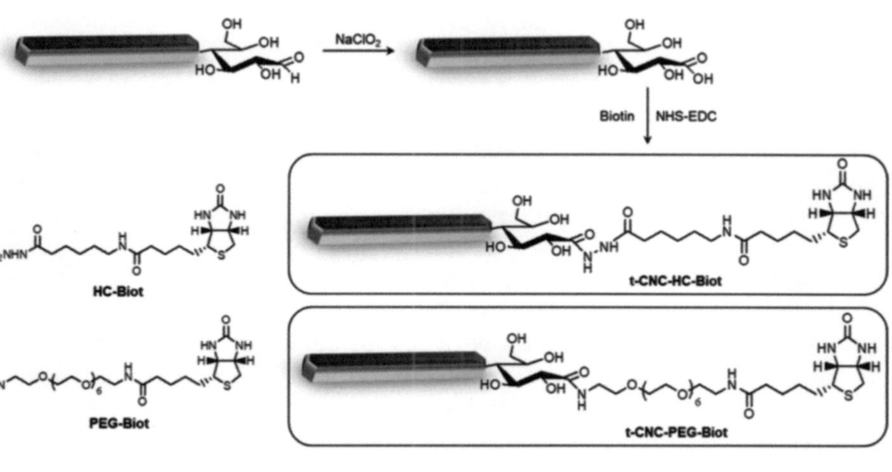

Figure 3.
Schematic illustration of carboxyamine condensation reaction at nanocellulose reducing end published in 2018 [27].

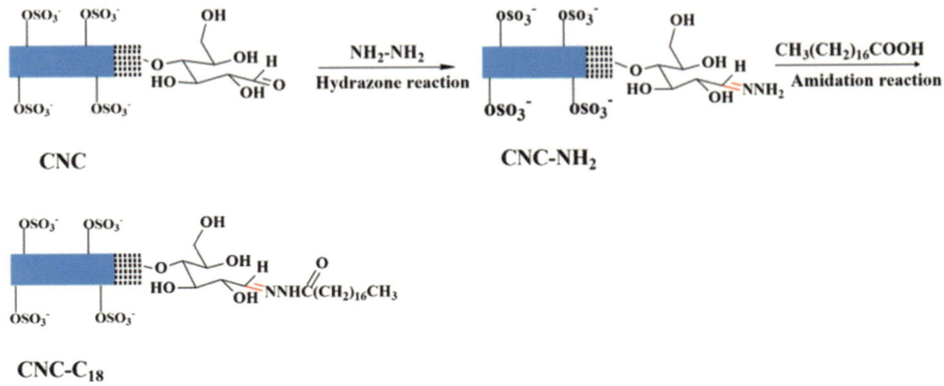

Figure 4.
Synthesis route of CNC-C18 via the hydrazone and amidation reaction published in 2017 [29].

pave a new way to endow unique characters or functional species on reducing end of cellulose based nanomaterials [30].

According to the literature, the motivation of CNC end modification is mainly focusing on the aldehyde group transformation to introduce new functional groups and polymer grafting wit functional species on the end of CNCs as shown in **Tables 1** and **2**. For the functional group derivazation by the carboxyamine condensation, aldimine condensation and hydrazone reaction as well as knoevenagel condensation reactions, acetyl group ($-COCH_3$), thiol group ($-SH$) and amino group ($-NH_2$) are introduced on the CNC reducing end, and the purpose of these derived functional groups are primary for the improvement of redispersion of CNCs, preparation of Pickering emulsions, template for nanomaterial preparation and conventional free radical polymerization according to **Table 1**. The radiolabeled molecule, fluorescent molecule and biotin derivatives are also grafted on the CNC reducing end to impart the CNCs with biological features for the applications in the biomedical fields, such as antimicrobial properties as shown in **Table 1**. For instance, the chlorotriazine reactive dye conjugated with CNCs produced an effective emulsifying effect of Pickering emulsions. The end of CNC chemically decorated with a hydrophobic dye species while maintaining hydroxyl functionality on CNC surface endowed it with amphiphilic property, thereby increasing its capacity for adsorption on the oil/water interface [45]. Except for the functional group derivazation on the CNC reducing end, polymer grafting reactions on the CNC end were also reported primary via atom transfer radical polymerization, carboxyamine condensation and aldimine condensation reactions to endow CNCs with stimuli-responsive features and apply it for stabilizers in Pickering emulsions. According to the report in the literature, the main grafting polymers are polystyrene, poly(N-isopropyl acrylamide, poly(sodium 4-vinylbenzene sulfonate) and poly(amidoamine) dendrimer as shown in **Table 2**. Compared to the conventional CNC surface grafting reaction, the well-preserved original structure of CNCs can be achieved, thereby offering an intact surface chemistry of resulted CNCs. For instance, temperature responsive polyetheramine polymers were grafted on the end of CNCs via a carboxyamine condensation reaction, and resulted polyetheramine grafted CNCs showed the temperature responsive property, which opens a new approach to design drug delivery carrier materials in biomedical applications.

Reaction Types	Functional Species	Functionality
carboxyamine condensation	thiol group (–SH)	adsorption on gold surfaces [31]
aldimine condensation	thiol group (–SH)	adsorption on silver surfaces [32]
carboxyamine condensation	thiol group (–SH)	dispersion property [33]
carboxyamine condensation	thiol group (–SH)	redispersion property [34]
aldimine condensation	thiol group (–SH)	redispersion property [34]
hydrazone reaction	amino group ($-NH_2$)	Pickering emulsions [29]
carboxyamine condensation	amino group ($-NH_2$)	radical polymerization [35]
carboxyamine condensation	azide group ($-N_3$)	templated nanobiomaterial [35]
knoevenagel condensation	acetyl group (OCH_3)	Self-assembly [30]
hydrazone reaction	fluorescent dye	biomedical application [26]
hydrazone reaction	radiolabled molecule	biomedical application [26]

Table 1.
Strategy of functional group transformations on the CNC end.

Reaction Types	Functional Polymers	Functionality
carboxyamine condensation	polyetheramine	responsive property [36, 37]
carboxyamine condensation	poly(amidoamine) dendrimers	responsive property [38]
carboxyamine condensation	poly(N-isopropylacrylamide)	self-assembly [39]
carboxyamine condensation	poly[2-(methacryloyloxy)ethyl]-trimethylammonium chloride	self-assembly [39]
carboxyamine condensation	poly(sodium 4-vinyl-benzene sulfonate)	self-assembly [39]
carboxyamine condensation	poly(amidoamine) dendrimer	supramolecular assembly [40]
carboxyamine condensation	poly(N-isopropylacrylamide)	responsive liquid crystals [41]
aldimine condensation	poly([2-(2-(methoxy ethoxy) ethoxy)- ethylacrylate]	liquid crystalline property [42]
aldimine condensation	poly(sodium 4-vinylbenzene sulfonate)	liquid crystalline property [43]
aldimine condensation	polystyrene	Pickering emulsion [44]

Table 2.
Strategy of grafted polymers on the CNC end.

3.2 Reducing end modification of CNCs on miscibility and mechanical performance of its nanocomposites

The application of CNC reducing end modification is still new according to the report in the literature. It is a promising approach to tailor the miscibility and mechanical performance of CNC-reinforced elastomer sustainable composite system as the aldehyde groups on CNC end are reacted with functional groups to transform its property, especially hydrophilic or hydrophobic character, while the well-preserved surface hydroxyl groups can be achieved, and its stable percolation networks enabled to be well-maintained, resulting in the composite performance enhancement. Taking recent published work from Prof. Lin and Alain's group in term of CNC end modification reinforced elastomer nanocomposites as a typical example, the excellent interfacial interaction among modified CNCs and styrene-butadiene-styrene (SBS) copolymer elastomer matrix and well-dispersion of reduced end modified CNCs were achieved combined with the UV light-initiated thiol-ene click reaction [46]. The CNC surface property and its stable percolation networks were well preserved in this novel strategy resulting in property improvement in term of mechanical performance compared to the previous reported CNC surface modification approach. In particular, the elongation-at-break values are almost no chance compared to the raw rubber materials. The similar work in term of reducing end modified CNC-reinforced natural rubber elastomer composite system was subsequent reported in the same group. The thiol group was first decorated on CNC end via the group transformation strategy named as aldimine condensation in **Table 1**, and the resulted thiol group modified CNCs was further reacted with natural rubber matrix via the UV light-initiated thiol-ene click reaction for construction of cross-linking network structures. Attributing to the cross-linking at the interface and reinforcement effect as well as the retaining stable percolating network in the CNC-natural rubber elastomer nanocomposites, the excellent interfacial miscibility and mechanical properties were achieved [47].

3.3 Economic analysis of CNCs reinforced elastomer nanocomposites

To give a comprehensive evaluation of competitive advantages by using CNC nanomaterials for reinforced elastomer composites, a simple economic analysis is conducted by comparative study of different types of fillers in reinforced elastomer composite materials. Carbon fibers, carbon nanotubes, and CNCs as representative fillers are comparatively studied attributed to their excellent elastic modulus. For instance, the elastic modulus of carbon fibers ranges from 150 to 500 GPa, while CNCs has elastic modulus between 110 and 220 GPa [48]. According to the literature, the cost of carbon fiber is \$20/kg, which is much lower than carbon nanotube at \$100/kg [49]. However, the prices of these carbon materials are much high for reinforced composite materials compared to the relatively less expensive CNC nanomaterial. Its price ranges from \$4 to 10/kg and expects to reach at \$5/kg or even less as the increasingly technique updates in terms of CNC preparation [50], so its price is around 4-times less than carbon fibers. Therefore, the CNCs has competitive advantages compared to carbon fibers and carbon nanotubes as it is important to choose reinforced fillers with excellent elastic modulus but also decent price for reinforced elastomer composites.

4. Conclusions

The progress in terms of CNC reducing end and surface modification to tailor miscibility and mechanical property of CNC-reinforced elastomer sustainable composites are summarized in this chapter. Compared to the traditional surface modification of CNCs, its unique traits of reducing end modification to establish the percolating network of CNC-reinforced elastomer composites still demands in-depth study. Reducing end modifications of CNCs tailoring PU, PDMS, polyacrylate and other types of rubber elastomeric materials are still no studied up-to-now except for reinforced styrene-butadiene-styrene copolymer elastomer and nature rubber composite system. As the original structure of CNC can be well-preserved for this novel end modification approach, it has wide applications in antibacterial, self-healing, recycling vitrimer, Pickering emulsions and chiral photonic smart material design in near future. The vitrimer as a new type of dynamic covalent cross-linked network polymers enables to achieve cross-linked CNC elastomer composites with recycling and reprocessing [51]. Due to its unique chiral liquid crystal structure, CNCs/poly(deep eutectic solvents) photonic elastomer composites displayed dynamic structural color and self-healing ability [52, 53].

Acknowledgements

We appreciate the financial support from the Natural Science Foundation of Hebei Province (Grant No. E2020203063) and Department of Education Foundation of Hebei Province (Grant No. QN2020104).

Author details

Jinlong Zhang[1*], Qinglin Wu[2*] and Weiguo Li[3]

1 School for Engineering of Matter, Transport and Energy, Arizona State University, Tempe, AZ, USA

2 School of Renewable Natural Resources, Louisiana State University Agricultural Center, Baton Rouge, LA, USA

3 College of Aerospace Engineering, Chongqing University, Chongqing, China

*Address all correspondence to: jzhan620@asu.edu and wuqing@lsu.edu

IntechOpen

© 2024 The Author(s). Licensee IntechOpen. This chapter is distributed under the terms of the Creative Commons Attribution License (http://creativecommons.org/licenses/by/3.0), which permits unrestricted use, distribution, and reproduction in any medium, provided the original work is properly cited.

References

[1] Science to enable sustainable plastics-A white paper from the 8th chemical science and society summit (CS3). 2020. Available from: rsc.li/sustainable-plastics-report

[2] Zhang JL, Wu QL, Zhang XQ, Gwon J, Zhang R, Negulescu I. Mechanical and water resistance performance of physically cross-linked poly(vinyl alcohol) composite with poly(dopamine) modified cellulose nanocrystals. Materials Today Communications. 2022;**33**:104413. DOI: 10.1016/j.mtcomm.2022.104413

[3] Mohanty AK, Vivekanandhan S, Pin JM, Misra M. Composites from renewable and sustainable resources: Challenges and innovations. Science. 2018;**6414**:536-542. DOI: 10.1126/science.aat9072

[4] Hillmyer MA. The promise of plastics from plants. Science. 2017;**6365**:868-870. DOI: 10.1126/science.aao6711

[5] Mi HY, Jing X, Peng J, Salick MR, Peng XF, Turng LS. Poly(e-caprolactone) (PCL)/cellulose nano-crystal (CNC) nanocomposites and foams. Cellulose. 2014;**21**:2727-2741. DOI: 10.1007/s10570-014-0327-y

[6] Wang N, Li ZK, Zhu G, Yu ZC, Qin RY, Zhou J, et al. Chemical modifications on cellulose nanocrystals for composites: Surface chemistry to tailored compatibility and mechanical enhancement. Macromolecules. 2023;**56**:7505-7519. DOI: 10.1021/acs.macromol. 3c00933

[7] Rosilo H, Kontturi E, Seitsonen J, Kolehmainen E, Ikkala O. Transition to reinforced state by percolating domains of intercalated brush-modified cellulose nanocrystals and poly(butadiene) in cross-linked composites based on thiol-ene click chemistry. Biomacromolecules. 2013;**14**:1547-1554. DOI: 10.1021/bm400185z

[8] Kanoth BP, Claudino M, Johansson M, Berglund LA, Zhou Q. Biocomposites from natural rubber: Synergistic effects of functionalized cellulose nanocrystals as both reinforcing and cross-linking agents via free-radical thiolene chemistry. ACS Applied Materials & Interfaces. 2015;**7**:16303-16310. DOI: 10.1021/acsami.5b03115

[9] Phua SL, Yang LP, Toh CL, Huang S, Tsakadze Z, Lau SK, et al. Reinforcement of polyether polyurethane with dopamine-modified clay: The role of interfacial hydrogen bonding. ACS Applied Materials & Interfaces. 2012;**4**:4571-4578. DOI: 10.1021/am3024405

[10] Sakakibara K, Moriki Y, Yano H, Tsujii Y. Strategy for the improvement of the mechanical properties of cellulose nanofiber-reinforced high-density polyethylene nanocomposites using diblock copolymer dispersants. ACS Applied Materials & Interfaces. 2017;**9**:44079-44087. DOI: 10.1021/acsami.7b13963

[11] Marais A, Kochumalayil JJ, Nilsson C, Fogelstrom L, Gamstedt EK. Toward an alternative compatibilizer for PLA/cellulose composites: Grafting of xyloglucan with PLA. Carbohydrate Polymers. 2012;**89**:1038-1043. DOI: 10.1016/j.carbpol.2012.03.051

[12] Rojas OJ, Montero GA, Habibi Y. Electrospun nanocomposites from polystyrene loaded with cellulose nanowhiskers. Journal of Applied

Polymer Science. 2009;**113**:927-935. DOI: 10.1002/app. 30011

[13] Batmaz R, Mohammed N, Zaman M, Minhas G, Berry RM, Tam KC. Cellulose nano-crystals as promising adsorbents for the removal of cationic dyes. Cellulose. 2014;**21**:1655-1665. DOI: 10.1007/s1 0570-014-0168-8

[14] Mariano M, Kissi NE, Dufresne A. Cellulose nanocrystals and related nanocomposites: Review of some properties and challenges. Journal of Polymer Science: Polymer Physics. 2014;**152**:791-806. DOI: 10.1002/polb.23490

[15] Mangalam AP, Simonsen J, Benight AS. Cellulose/DNA hydid nanomaterials. Biomacromolecules. 2009;**10**:497-504. DOI: 10.1186/s13036-019-0191-2

[16] Habibi Y, Lucia LA, Rojas OJ. Cellulose nanocrystals: Chemistry, self-assembly, and applications. Chemical Reviews. 2010;**110**:3479-3500. DOI: 10.1021/cr900339w

[17] Habibi Y. Key advances in the chemical modification of nanocelluloses. Chemical Society Reviews. 2014;**43**:1519-1542. DOI: 10.1039/C3CS60204D

[18] Ghasemlou M, Daver F, Ivanova EP, Habibi Y, Adhikari B. Surface modifications of nanocellulose: From synthesis to high-performance nanocomposites. Progress in Polymer Science. 2021;**119**:101418. DOI: 10.1016/j.progpolymsci.2021.101418

[19] Favier V, Canova GR, Cavaillé JY, Chanzy H, Dufresne A, Gauthier C. Nanocomposite materials from latex and cellulose whiskers. Polymers for Advanced Technology. 1995;**6**:351-355. DOI: 10.1002/pat.1995. 220060514

[20] Dastjerdi Z, Cranston ED, Dubé MA. Pressure sensitive adhesive property modification using cellulose nanocrystals. International Journal of Adhesion and Adhesives. 2018;**81**:36-42. DOI: 10.1016/j.ijadhadh.2017.11.009

[21] Trovatti E, Oliveira L, Freire CSR, Silvestre AJD, Neto CP, Pinto JJCC, et al. Novel bacterial cellulose-acrylic resin nanocomposites. Composites Science and Technology. 2010;**70**:1148-1153. DOI: 10.1016/j.comp scitech.2010.02.031

[22] Lacerda PSS, Barros-Timmons AMMV, Freire CSR, Silvestre AJD, Neto CP. Nanostructured composites obtained by ATRP sleeving of bacterial cellulose nanofibers with acrylate polymers. Biomacromolecules. 2013;**14**:2063-2073. DOI: 10.1021/bm400432b

[23] Panganiban B, Qiao BF, Jiang T, DelRe C, Obadia MM, Nguyen TD, et al. Random heteropolymers preserve protein function in foreign environments. Science. 2018;**359**:1239-1243. DOI: 10.1126/science.aao0335

[24] Zhang JL, Li MC, Zhang XQ, Ren SX, Dong LL, Lee SY, et al. Water redispersion and cytotoxicity of reducing end-modified cellulose nanocrystals by grafting long-chain poly(ethylene oxide). International Journal of Biological Macromolecules. 2021;**180**:143-151. DOI: 10.1007/s10570-019-02379-z

[25] Chemin M, Moreau C, Cathala B, Villares A. Divergent growth of poly(amidoamine) dendrimer-like branched polymers at the reducing end of cellulose nanocrystals. Carbohydrate Polymers. 2022;**279**:1198008. DOI: 10.1016/j.carbpol.2021.119008

[26] Imlimthan S, Otaru S, Keinanen O, Correia A, Lintinen K, Santos HA, et al. Radiolabeled molecular imaging probes

for the in vivo evaluation of cellulose nanocrystals for biomedical applications. Biomacromolecules. 2019;**20**:674-683. DOI: 10.1021/acs.biomac.8b01313

[27] Villares A, Moreau C, Cathala B. Star-like supramolecular complexes of reducing-end-functionalized cellulose nanocrystals. ACS Omega. 2018;**3**:16203-16211. DOI: 10.1021/acsomega.8b02559

[28] Tan T, Dufresne A, Zhu G, Lin N. Regioselective modification at reducing end aldehydes of cellulose nanocrystals and mercerization. ACS Sustainable Chemistry & Engineering. 2023;**11**:4485-4497. DOI: 10.1021/acssuschemeng.2c07713

[29] Du WB, Guo J, Li HM, Gao Y. Heterogeneously modified cellulose nanocrystals-stabilized Pickering emulsion: Preparation and their template application for the creation of PS microspheres with amino-rich surfaces. ACS Sustainable Chemistry & Engineering. 2017;**5**:7514-7523. DOI: 10.1021/acssuschemeng.7b00375

[30] Heise K, Koso T, Pitkanen L, Potthast A, AWT K, Kostiainen MA, et al. Knoevenagel condensation for modifying the reducing end groups of cellulose nanocrystals. ACS Macro Letters. 2019;**8**:1642-1647. DOI: 10.1021/acsmacrolett.9b00838

[31] Arcot LR, Lundahl M, Rojas OJ, Laine L. Asymmetric cellulose nanocrystals: Thiolation of reducing end groups via NHS-EDC coupling. Cellulose. 2014;**21**:4209-4218. DOI: 10.1007/s10570-014-0426-9

[32] Lokanathan AR, Nykanen A, Seitsonen J, Johansson LS, Campbell J, Rojas OJ, et al. Cilia-mimetic hairy surfaces based on end-immobilized nanocellulose colloidal rods. Biomacromolecules. 2013;**14**:2807-2813. DOI: 10.1021/bm400633r

[33] Zhao XP, Chen ZY, Lin Y, Ma JZ. Water redispersion and cytotoxicity of reducing end-modified cellulose nanocrystals by grafting long-chain poly(ethylene oxide). International Journal of Biological Macromolecules. 2021;**180**:143-151. DOI: 10.1016/j.ijbiomac.2021.03.010

[34] Li L, Tao H, Wu BL, Zhu G, Li K, Lin Y. Triazole end-grafting on cellulose nanocrystals for waterRedispersion improvement and reactive enhancement to nanocomposites. ACS Sustainable Chemistry & Engineering. 2018;**6**:14888-14900. DOI: 10.1021/acssuschemeng.8b03407

[35] Sipahi-Saglam E, Gelbrich M, Gruber E. Topochemically modified cellulose. Cellulose. 2003;**10**:237-250. DOI: 10.1023/A: 10251 51701985

[36] Lin FB, Cousin F, Putaux JL, Jean B. Temperature-controlled star-shaped cellulose nanocrystal assemblies resulting from asymmetric polymer grafting. ACS Macro Letters. 2019;**8**:345-351. DOI: 10.1021/acsmacrolett.8b01005

[37] Lin FB, Pignon F, Putaux JL, Jean B. Temperature-triggered formation of a cellulose II nanocrystal network through regioselective derivatization. Nanoscale. 2021;**13**:6447-6460. DOI: 10.1039/D0NR08597A

[38] Chemin M, Moreau C, Cathala B, Villares A. Asymmetric modification of cellulose nanocrystals with PAMAM dendrimers for the preparation of pH-responsive hairy surfaces. Carbohydrate Polymers. 2020;**249**:116779. DOI: 10.1016/j.carbpol.2020.116779

[39] Zoppe JO, Dupire AVM, Lachat TGG, Lemal P, Rodriguez-Lorenzo L, Petri-Fink A, et al. Cellulose nanocrystals with tethered polymer

chains: Chemically patchy versus uniform decoration. ACS Macro Letters. 2017;**6**:892-897

[40] Chemin M, Moreau C, Cathala B, Villares A. Divergent growth of poly(amidoamine) dendrimer-like branched polymers at the reducing end of cellulose nanocrystals. Carbohydrate Polymers. 2022;**279**:119008. DOI: 10.1021/acsmacrolett.7b00383

[41] Risteen B, Delepierre G, Srinivasarao M, Weder C, Russo P, Reichmanis E, et al. Thermally switchable liquid crystals based on cellulose nanocrystals with patchy polymer grafts. Small. 2018;**14**:1802060. DOI: 10.1002/smll.201870218

[42] Delepierre G, Traeger H, Adamcik J, Cranston ED, Weder C, Zoppe JO. Liquid crystalline properties of symmetric and asymmetric end grafted cellulose nanocrystals. Biomacromolecules. 2021;**22**:3552-3564. DOI: 10.1021/acs.biomac.1c00644

[43] Delepierre G, Heise K, Malinen K, Koso T, Pitkänen L, Cranston ED, et al. Challenges in synthesis and analysis of asymmetrically grafted cellulose nanocrystals via atom transfer radical polymerization. Biomacromolecules. 2021;**22**:2702-2717. DOI: 10.1021/acs.biomac. 1c00392

[44] Tang CX, Spinney S, Shi ZQ, Tang JT, Peng BL, Luo JH, et al. Amphiphilic cellulose nanocrystals for enhanced Pickering emulsion stabilization. Langmuir. 2018;**34**:12897-12905. DOI: 10.1021/acs.langmuir.8b02437

[45] Surov OV, Voronova MI, Zakharov AG. Synthesis and properties of cellulose nanocrystal conjugates with reactive dyes. Cellulose. 2021;**28**:6269-6285. DOI: 10.1007/s10570-021-03898-4

[46] Tao H, Dufresne A, Lin N. Double-network formation and mechanical enhancement of reducing end-modified cellulose nanocrystals to the thermoplastic elastomer based on click reaction and bulk cross-linking. Macromolecules. 2019;**52**:5894-5906. DOI: 10.1021/acs.macromol.9b01213

[47] Zhu G, Dufresne A. Synergistic reinforcing and cross-linking effect of thiol-ene-modified cellulose nanofibrils on natural rubber. Carbohydrate Polymers. 2022;**278**:118954. DOI: 10.1016/j.carbpol. 2021.118954

[48] Moon RJ, Martini A, Nairn J, Simonsen J, Youngblood J. Cellulose nanomaterials review: Structure, properties and nanocomposites. Chemical Society Reviews. 2011;**40**:3941-3994. DOI: 10.1039/C0CS00108B

[49] Huang J, Rodrigue D. Comparison of the mechanical properties between carbon nanotube and nanocrystalline cellulose pol ypropylene based nano-composites. Materials & Design. 2015;**65**:974-982. DOI: 10.1016/j.matdes.2014.09.081

[50] Norizan MN, Shazleen SS, Alias AH, Sabaruddin FA, Asyraf MRM, Zainudin ES, et al. Nanocellulose-based nanocomposites for sustainable applications: A review. Nanomaterials. 2022;**12**:3483. DOI: 10.3390/nano12193483

[51] Zhang JL, Li WG, Wu QL. Renewable resource-derived elastomer vitrimer and its sustainable manufacturing and application in extreme environmental conditions. BioResources. 2023;**18**:4395-4398. DOI: 10.15376/biores.18.3.4395-4398

[52] Li XK, Liu JZ, Guo QQ, Zhang XX, Tian M. Polymerizable deep eutectic solvent-based skin-like elastomers with

dynamic schemochrome and self-healing ability. Small. 2022;**18**:2201012. DOI: 10.1002/smll.202201012

[53] Mota-Morales JD, Morales-Narváez ED. Transforming nature into the next generation of bio-based flexible devices: New avenues using deep eutectic systems. Matter. 2021;**4**:2141-2162. DOI: 10.1016/j.matt.2021.05.009